Glim: An Introduction

M. J. R. Healy

London School of Hygiene and Tropical Medicine

CLARENDON PRESS · OXFORD · 1988

Oxford University Press, Walton Street, Oxford OX2 6DP

Oxford New York Toronto
Delhi Bombay Calcutta Madras Karachi
Petaling Jaya Singapore Hong Kong Tokyo
Nairobi Dar es Salaam Cape Town
Melbourne Auckland
and associated companies in
Beirut Berlin Ibadan Nicosia

Oxford is a trade mark of Oxford University Press

Published in the United States
by Oxford University Press, New York

British Library Cataloguing in Publication Data

Healy, M. J. R.
GLIM: an introduction.
1. GLIM (Computer programs)
I. Title
519.5′028′553   QA276.4
ISBN 0-19-852213-4

Library of Congress Cataloging in Publication Data
Healy, M. J. R.
GLIM: an introduction.
Includes index.
1. GLIM (Computer program)   2. Linear models
(Statistics)—Data processing.   I. Title.
QA279.H43   1988   519.5′028′55369   87-24816
ISBN 0-19-852213-4

Typeset by H. Charlesworth & Co. Ltd., Huddersfield
Printed in Great Britain
at the University Printing House, Oxford
by David Stanford
Printer to the University

Glim: An Introduction

To FY
With thanks

Contents

0
Introducing Glim

Linear modelling is one of the most basic of statistical operations, comprising as it does the bulk of least squares and analysis of variance theory and practice. The generalized linear models introduced by Nelder and Wedderburn in 1972 greatly extend the scope of the methodology, pulling together such topics as contingency tables and probit analysis which had been regarded as requiring special methods of their own.

The calculations needed to fit a linear model can be fairly heavy, the more so when iterative methods are required. As a result, many computer programs have been written for performing them and linear model modules are included in all the standard computer packages. Many of these are unnecessarily limited in scope and few take account of the 1972 generalizations. A Royal Statistical Society group headed by J. A. Nelder constructed a special-purpose package called Glim (Generalised Linear Interactive Modelling) which was primarily intended to facilitate the fitting and investigation of generalized linear models. This has grown into a specialized computer language which in many respects meets the needs of the practical data analyst better than any of its competitors.

Glim's very power means that the full set of facilities described in its manual is rather daunting to the beginner. In addition, the output from the package is not always easy to understand. The purpose of this book is to introduce its facilities and to explain their use without trying to explore all the possibilities available—new uses for Glim are still being discovered and these are described in the *Glim Newsletter* which appears regularly. I assume that you, the reader, have some acquaintance with statistical methods in general and with linear modelling (at least as far as multiple regression) in particular. The latter is a large topic indeed and can be studied using one of the many textbooks on the market.

At the time of writing, Glim is widely available in two versions. These are officially releases 3.12 and 3.77, but I shall refer to them for brevity as Glim3 and Glim77 respectively. Both can be run on most mainframe and minicomputers. Glim3 runs on 8-bit micros under the CP/M operating system, and Glim77 runs on 16-bit machines under MS-DOS or PC-DOS. Other implementations are also available. Glim also forms part of the Genstat package. This book is designed to be usable with either of the two versions.

Glim is maintained and promulgated by the Numerical Algorithms Group with the support of the Royal Statistical Society. Enquiries should be addressed to:

The Glim Coordinator
NAG Central Office
256 Banbury Road
Oxford OX2 7DE
UK

or to

NAG Inc
1101 31st St, Suite 100
Downers Grove
Illinois 60515–1263
USA

The manuals for Glim3 and Glim77 and the *Glim Newsletter* can be obtained from the same addresses.

The theory underlying Glim is given rather succinctly in the original article (J. A. Nelder and R. W. M. Wedderburn (1972) Generalised linear models. *J. R. statist. Soc. A* **135**, 378–84) and is also outlined in the manuals. A much fuller account which describes many different applications of the package is the book *Generalised Linear Models* by P. McCullagh and J. A. Nelder (1983) published by Chapman and Hall.

There are many textbooks on the statistics of linear modelling. Some popular ones are:

N. R. Draper and H. Smith (1981) *Applied regression analysis* (2nd edn). Wiley, New York.

S. R. Searle (1971) *Linear models.* Wiley, New York.

S. Weisburg (1980) *Applied linear regression.* Wiley, New York.

On more specialized topics, see for example:

R. D. Cook and S. Weisburg (1982) *Residuals and influence in regression.* Chapman and Hall, London.

A. C. Atkinson (1985) *Plots, transformations, and regression.* Oxford University Press, Oxford.

J. L. Fleiss (1981) *Statistical methods for rates and proportions.* Wiley, New York.

J. D. Kalbfleisch and R. L. Prentice (1980) *The statistical analysis of failure time data.* Wiley, New York.

Y. M. M. Bishop, S. E. Fienburg, and P. W. Holland (1978) *Discrete multivariate analysis: theory and practice.* MIT Press, Cambridge, MA.

Most of the examples in this book are based on real data taken from published material. I have not acknowledged the authors individually since I have usually altered the figures to meet my special purposes, but I am grateful to them none the less. I am also grateful to the Imperial Cancer Research Fund for assistance in preparing the text.

1

Data input and scrutiny

1.1 Linear models

Suppose we have some data relating to a random quantity y whose mean value or *expectation* we write as $E(y)$ and that this mean value depends upon one or more other quantities which we denote by x_1, x_2, ..., x_p. In the simplest situation the relation between $E(y)$ and the xs will be *linear* so that we can write

$$E(y) = \beta_0 + \beta_1 x_1 + \cdots + \beta_p x_p$$

where the βs are a set of coefficients whose values we would like to know. This is part of a *linear model* for the way in which the statistical behaviour of y is related to the values of the xs. To complete the model we need to describe the way in which the actual values of y vary about their mean, in fact to describe the *distribution* of the ys at any particular set of values of the xs. We might for instance specify that the variance of this distribution is a constant quantity σ^2 for any fixed x-values. For some purposes we may need to go further and specify that the distribution is Normal. This simple form of model is usually met with first in statistics under the name of *multiple regression*. The quantity y is sometimes called the *dependent variate*, while the xs are the *independent variates* (a doubly misleading name since they need be neither independent nor variates), the *predictors*, or the *carriers*. The βs are the *regression coefficients* and σ is the *residual standard deviation*. Later on we shall greatly widen the scope of our methods by investigating linear models which describe some mathematical function of y rather than y itself and by making different assumptions about the distribution of the ys about their mean (the *error distribution*).

Given a body of data providing values of y and the corresponding xs we *fit* the model by forming *estimates* of its *parameters*. These include the βs and any quantities we may need to specify the error distribution—if this is Normal, all we need is the variance σ^2. To do this we shall use the method of *maximum likelihood*—crudely speaking, we adopt as estimates of the parameters those values for which the observed data are most probable. These estimates are known from statistical theory to have various nice properties; in large samples they are correct on average and are as precise as possible given the data available. If the error distribution is Normal, this method is the same as the method of *least squares* which picks as estimates parameter values which minimize the sum of squares of the *residuals*, the differences

3

between the y-values predicted from the model and those actually observed.

1.2 The Glim package

Glim (standing for Generalized Linear Interactive Modelling) is a computer program for fitting linear models and the more general models relating to them. Glim3, the third release of the program, has been extensively used on machines ranging from 64k CP/M micros to Cray 1 supercomputers, and a new version Glim77 is already widely available. In this book I shall mark facilities which are peculiar to Glim77 and not available in Glim3 by @ signs. The two versions produce slightly different output. The examples shown in this book will use the Glim77 output, but users of Glim3 should not find problems in interpreting their versions.

I am going to assume that you, the reader, have access to a version of Glim and that you know how to persuade your computer to load and enter the program. I shall also assume that the program is being used interactively so that you are looking at a screen on which appears some kind of message which might read

```
GLIM 3.77 update 0 (copyright)1985 Royal Statistical Society, London
```

probably followed at the start of the next line by some kind of *prompt* such as > > or ?, indicating that Glim is waiting for you to give it a *command*. Glim can also be used in batch mode, when a whole session is planned in advance and presented to the computer in a single chunk, but the program is much easier to learn and to use interactively, especially if your computer can be persuaded to provide for you a printed record of what has being going on for future reference. Far the best way to use this book is to study it while sitting at a keyboard.

@@@@
Some versions of Glim77 will ask you before anything else happens whether you want a *transcript file* (other versions demand some action on your part before the program is entered). If you reply with a file name, anything that appears on the screen during your session will be recorded in the file. This is an immensely useful feature, since you can inspect the file at leisure, perhaps edit out the unnecessary bits, and have it printed out as a hard copy record.

Every line in the transcript file will be identified by a code letter which indicates its source. The letters are
 i for Input
 v for Verification of macros (see Section 9.7)
 w for Warning messages
 f for Fault or error messages
 h for Help messages
 o for Ordinary output from the program
@@@@

Glim is best thought of as a special-purpose *computer language* in which you can specify the operations you wish to have performed. A Glim session consists in the user issuing a sequence of *directives*, most of which take the form of *commands* instructing the program to do something. In what follows I shall introduce the directives and explain how they are used. There is a complete list in Appendix A.

Every Glim directive starts with a special symbol which the program recognizes. In almost all implementations this is a dollar sign $ (occasionally a pound sign £ is used). Glim uses a few more symbols with special meanings and these do sometimes differ from one machine to another. As our first Glim directive we might try typing

$ENVIRONMENT I

where the letter I stands for 'Installation details'. Note that Glim3 recognizes only capital letters—I shall use capitals in the examples although the Glim77 user can equally well use small letters. You must always finish the typed line by hitting the RETURN key on your keyboard; the computer cannot 'see' it until you have done this.

Glim will now obey the command and give you some output. The details of what you will get depends upon just what your machine provides—this is after all what we are trying to find out. On the machine I am using at present I get

```
Installation:   2060 implementation,  KGA & P.A.G.
  Release 3.77 of GLIM  -  Update  0
      directive symbol is  $
      repetition symbol is  :
      substitution symbol is  #
      quote symbol is  '
      function symbol is  %
      end-of-record symbol is  !
      largest integer is  34359738359
```

These special symbols are the same as those used in the official Glim manuals and I shall be using them throughout this book. Do not bother for the moment about what the other special symbols mean; I will explain them as we get to them.

$ENVIRONMENT is quite a bit of typing for a fairly modest return and most of it can be avoided. In directives and elsewhere, *Glim only attends to the first four characters, throwing away the rest*. Thus you would get the same effect by typing

$ENV I

Many directives can actually be shortened even more; the minimum abbreviations are given in Appendix A but they are a nuisance to remember

($P, $PA, and $PAG stand for three quite different things) and it is easiest to settle for four characters or more as standard practice. In this book I shall use the full directive names to make the examples easier to read. When you try them on your machine you can use the shortened forms.

Another useful directive we can introduce at this stage is

$C

which is followed by a *comment*, any string of characters which appear after the C and before the next $ sign. This string is totally ignored by the program but it will appear on any printed output that your session produces. A few judicious comments are essential if you are going to be able to distinguish between yesterday's Glim output and that of the previous week. Another way of introducing a comment is to use the *end-of-record* symbol !. Glim will ignore anything typed after this on the same line.

Some Glim directives have to be *terminated* by a $ sign before Glim will obey them. This can be the start of the next directive, but if nothing seems to be happening it is always worth typing a $ (followed by RETURN). Glim77 will remind you that it is expecting something more by adding the directive name to the prompt.

1.3 Data input

The first task in a Glim session is likely to be the input of the data for the problem at hand. These will best be thought of as making up a *data matrix*, a two-way table of numbers with each column representing a *variable* and each row a *unit, case, subject, point,* or whatever. A small data matrix is shown in Table 1.3.1. Here we have three variables (height, weight, and chest circumference) measured on 10 subjects. In Glim terminology we have 10 *units* in our data and we need a directive to convey this information to the program—type

$UNITS 10

Next we must specify the variables in the problem—Glim calls them *vectors*. Unlike the units, we are going to need to refer to these one at a time, so we must give each one of them a name. We could use HEIGHT, WEIGHT, and CHEST provided we remember that Glim will throw away all but the first four letters of each name.

Suppose we decide to input the data a row at a time. As Glim reads in the successive numbers it must allocate them in sequence to HEIGHT, WEIGHT, CHEST, HEIGHT, WEIGHT, ... until a total of $10 \times 3 = 30$ numbers have been read in. We convey this further information by typing

$DATA HEIGHT WEIGHT CHEST

(Note that the names are merely separated by spaces. The format of Glim

Table 1.3.1 A data matrix

Height (cm)	Weight (kg)	Chest circ. (cm)
167.9	71.8	30.0
183.8	75.1	29.4
172.9	58.0	26.1
175.5	58.4	25.7
176.4	67.7	27.9
168.5	75.2	31.7
178.0	67.3	27.4
178.0	71.3	29.0
175.4	75.9	30.3
171.2	65.3	29.4

directives is almost entirely free, but you must not split a name between one line and the next.) Because of the $UNITS directive, Glim already knows that each vector consists of 10 items, or in Glim terminology that each vector has *length* 10.

Names provided by the user for Glim must be made up of letters and/or digits, starting with a letter. You must not use names such as VAR/1 or 3DATA. Because only the first four characters are kept, avoid using names like VAR1, VAR2, ...; VAR1 and VAR10 are the same name as far as Glim is concerned.

@@@@
Glim77 allows the underline symbol to be part of a name, so that names such as p_1 or H_IJ are permitted. It is wise not to use underline as the fourth character of a name, for reasons explained in Section 9.3. Note that Glim77, although it accepts lower-case letters, treats (for example) HEIG, Heig, and heig as all the same name.
@@@@

Now we are ready to do the actual data input. Type

$READ

followed by the data items row by row. The numbers can be typed in any old layout with spaces and RETURNs between them, but it is highly advisable to type them tidily in columns so that they can be quickly checked by eye. Do not split a number between two rows, and do not use commas in large numbers. Use a dot, not a comma, for a decimal point.

Glim is expecting 30 numbers and 30 numbers it insists on having. If you type too many or too few before the next $ sign, you will get an *error message*, probably saying 'DIRECTIVE EXPECTED BUT NOT FOUND' if you try to type too many numbers, or 'FAULT IN REAL NUMBER' if you try to type too few. Each of these will be followed by further explanation and it should be fairly clear what has happened. You will have to issue a new $READ directive and type in the 30 numbers all over again.

We could of course read in the data a column at a time if this were more convenient. This would need three $READ directives, each one preceded by its own $DATA directive, thus:

```
$DATA HEIGHT      $READ
167.9 183.8 172.9 175.5 176.4 168.5 178.0 178.0 175.4 171.2
$DATA WEIGHT      $READ
71.8 75.1 58.0 58.4 67.7 75.2 67.3 71.3 75.9 65.3
$DATA CHEST       $READ
30.0 29.4 26.1 25.7 27.9 31.7 27.4 29.0 30.3 29.4
```

@@@@

Glim77 provides a simpler way of putting values into a single vector, using the $ASSIGN directive. We could for example input a set of four numerical multipliers to a vector called MULT by typing

$ASSIGN MULT = -3, -1, 1, 3 $

Note the commas; also the use of the $ symbol by itself to show Glim that the list of numbers is finished. The vector MULT will now have length 4, different from the standard length of 10 set by the previous $UNITS directive. The $ASSIGN list can contain names as well as numbers. If we go on and type

$ASSIGN MULT = -5, MULT, 5 $

the vector MULT will now have length 6 and will contain the numbers -5, -3, -1, 1, 3, 5.

@@@@

1.4 Input from disk files

If you type in your data during a Glim session as described in the previous section you are liable to lose them at the end of the session, and this is a great nuisance if (as is usually the case) you need to return to them later for some further analyses. It is possible to *dump* data to a file and recover them later (see Section 8.5) but there is a lot to be said for starting out by making a disk file of your data using your local editor program. Very importantly, this enables you to scrutinize the data and correct any errors—this should be an essential first step before using the data for any statistical calculations. You can read in the contents of the file by issuing the appropriate $UNITS and $DATA commands and then typing

$DINPUT

instead of $READ.

There is one snag. Clearly Glim must be told the name of the file you want it to read from. This has to be done by following $DINPUT by the number of the *channel* or *logical unit* to which the file is attached. This is a Fortran concept (Glim is written in Fortran). Unfortunately connecting a named file to a channel is a highly installation-dependent matter; if you are lucky, you will be able to type in the file name in response to a question, but it may be necessary to do something at the time the Glim program is loaded and entered. What is more, some channel numbers will have been pre-empted by your terminal (and perhaps by a printer). You can get some information about these by typing

$ENVIRONMENT C

where the C stands for Channel. Further than this, you will have to consult a knowledgeable adviser. In the examples in this book I shall use channel 8; you may have to modify this.

1.5 Formatted input

This section can be omitted at a first reading.

Data are usually presented to Glim in *free format*. This means that the numbers must simply be presented in the right sequence to match the $DATA directive currently in force, separated by spaces or newlines and not split across lines or divided by commas, but without any special layout on the page. Occasionally a file will contain data in *fixed format* with the numbers divided into records (perhaps with no spaces between them) and the split into individual items being decided by the positions of the characters within the records. If this type of input is required, the $READ or $DINPUT directives must be preceded by a $FORMAT directive. This must be immediately followed by a Fortran-style format specification, including the parentheses but excluding the word FORMAT, standing on a line by itself (readers not familiar with Fortran format statements are referred to Appendix B). Note that the data items must be specified as reals (F or E format) in Fortran terminology. A $FORMAT directive remains in force for all subsequent $READs and $DINPUTs until a new one is issued or until the end of a *job* (see Section 1.8). To revert to free format, type

$FORMAT
FREE

The word FREE must appear at the beginning of a line.

Fixed format input is useful in the rare cases when data items are not separated by spaces. It can also be used (with Xs in the specification) to skip over certain fields which are not wanted in the analysis.

@@@@
Glim77 is more forgiving about the layout of the $FORMAT directive; the

Fortran format can start on the same line as the directive and the opening and closing parentheses can be omitted. The facilities of Fortran77 can be used. The directive

$FORMAT $

is taken to indicate free format.
@@@@

1.6 Looking at the data

Once you have input the data you may want to look at certain items, perhaps to check that they have been entered correctly or just to get a printed record. You can do this by using the $LOOK directive, following it by the names of the vectors you want to look at. With our data matrix the command

$LOOK HEIGHT WEIGHT

will produce the following output:

```
       HEIG    WEIG
 1    167.9   71.80
 2    183.8   75.10
 3    172.9   58.00
 4    175.5   58.40
 5    176.4   67.70
 6    168.5   75.20
 7    178.0   67.30
 8    178.0   71.30
 9    175.4   75.90
10    171.2   65.30
```

If necessary you can $LOOK at selected items from one or more vectors. Suppose you want to see the values of CHEST for the 3rd to 5th rows of the data matrix. You should type

$LOOK 3 5 CHEST $

and this will produce

```
3    26.10
4    25.70
5    27.90
```

When you look at your data you may well need to change certain values, either because they were entered wrongly or so as to see what a change in the data does to the results of the analysis. The $EDIT directive is used for this purpose. In its simplest form we could ask for item 3 in the CHEST vector to be changed to the value 36.0 by typing

$EDIT 3 CHEST 36.0 $

It is also possible to specify a range of item numbers provided that a sufficient number of data values are also included.

The $EDIT directive is handy for making temporary changes in the data. However, if the data are in a disk file, changes which are intended to be permanent (such as corrections to wrongly typed entries) should be made in the file outside Glim using an editor program.

1.7 Plotting

An extremely useful feature of Glim is the possibility of obtaining rather crude plots. If X and Y are two vectors, the command

$PLOT Y X $

will produce a plot on the screen of the values of Y (vertically) against those of X (horizontally). Single points are represented by the first letter of the y-variable name (Y in this example). If two or more points coincide a digit is printed, with 9 meaning '9 or more'. With our data the command

$PLOT WEIGHT HEIGHT $

produces the plot in Fig. 1.7.1. You can control the size of the plot by altering the width and height of the output channel—this obscure remark will be clarified in Section 8.4.

By putting several vector names in the directive it is possible to plot a number of different y-variables against the same x-variable. For example the command

$PLOT WEIGHT CHEST HEIGHT $

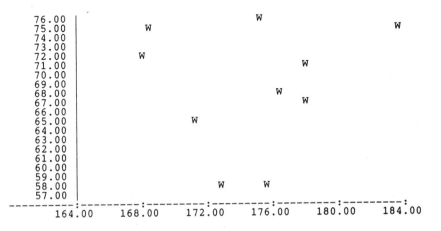

Fig. 1.7.1 Plot of weight against height, data from Table 1.3.1

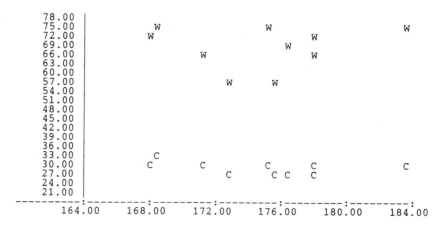

Fig. 1.7.2 Plot of weight and chest circumference against height, data from Table 1.3.1

gives the plot in Fig. 1.7.2. This of course only makes sense when the different *y*-variables can reasonably be plotted on the same scale. Note that in Glim3 variables with the same initial letter will be indistinguishable on the plot.

@@@@

Glim77 provides a means of specifying directly the symbols to be used in the plot. The required symbols for the successive *y*-variables are placed between single quotes at the end of the directive. Type

```
$PLOT WEIGHT CHEST HEIGHT 'pq'    $
```

and compare the result with Fig. 1.7.2 (note that Glim77 is quite happy with lower-case letters).

Glim77 also permits a number of *options* to be specified in brackets before the list of variables. You can for example type

```
$PLOT (ylimit=y1,y2  xlimit=x1,x2) YVAR XVAR $
```

where y1, y2 and x1, x2 are pairs of numbers giving the minimum and maximum plotting limits vertically and horizontally. Values outside the limits are not plotted. This is useful when several plots are required to be the same size. You can also control the size of the plot by including in the list of options

```
rows=r   cols=c
```

to specify that r rows and c columns are to be plotted (at least 2 rows and 21 columns will always be used).

Suppose we have a variable GRP whose values are integers 1, 2, ... By specifying the plotting symbols and naming GRP at the end of the directive, you can arrange that the points with different values in GRP are given different symbols on the plot. The number of different values in GRP must be specified by defining it to be a *factor* (see Section 2.4). Note that if GRP contains g different values and there are n_y y-variables to plot, $g \times n_y$ symbols must be specified. In our example suppose the vector AGE contained 1s for men under 25 and 2s for men 25 and over. We might type

```
$FACTOR AGE 2
$PLOT WEIGHT CHEST HEIGHT 'ABab' AGE $
```

The symbols A and B will be used for WEIGHT, a and b for CHEST. Users of the Minitab package will recognize the 'letter-plot' facility.

@@@@

@@@@

1.8 Histograms

Glim77 has a command which plots the histogram of the values in a vector. If the variable's name is X you simply have to type

```
$HISTOGRAM X $
```

The first character of the name will be used as the plotting symbol. As an example try typing

```
$UNITS 24    $DATA X    $READ
27 33 26 54 82 29 11 12 14 31 12
49 22 16 58 28 16 20 37 19 21 48 13
$HISTOGRAM X $
```

and you will get

```
[10.,30.) 16 XXXXXXXXXXXXXXXX
[30.,50.)  5 XXXXX
[50.,70.)  2 XX
[70.,90.]  1 X
```

The two sorts of bracket indicate that 30.0000 is included in the second histogram bar but 50.0000 is not.

This simple command can be extended in several ways. If you provide more than one variable name, a set of 'interleaved' histograms will be produced. You could for example put male and female values of some quantity into separate vectors and histogram them side by side. You can specify alternative plotting symbols by placing them in quotes at the end of the command, just as with the $PLOT command. In the same way, the name of a vector containing small integers which has been defined to be a factor can be given following the plotting symbols to give interleaved histograms corresponding to the different integer values. If our X quantities were accompanied by a second vector SEX having 1s for females and 2s for males, you could type

```
$FACTOR SEX 2
$HISTOGRAM X 'FM' SEX $
```

to plot male and female bars side by side.

It is possible to associate a vector of *weights* with each of the variables in a $HISTOGRAM directive. Normally, if there are n values of a variable X in a particular histogram bar the number of symbols plotted will be equal to n (or to a simple fraction of it if the full number is too large to fit on the page). If a weight variable W has been associated with X, the number of plotted symbols is based instead on the sum of the weights belonging to the appropriate values of X. The commonest use of this facility involves a weight vector whose items are 0s and 1s. Values of X with weight 1 are treated normally while those with weight 0 are simply ignored. This is a convenient way of treating certain data values as if they were missing; we shall meet it again.

The directive for a weighted histogram of the values in X using weights in W is

```
$HISTOGRAM X/W $
```

If several variables are included in the command, separate weight vectors can be specified for each one.

Left to itself, Glim will choose its own values for the number of histogram bars and the size of the plot. You can exert some control over these aspects of the command by specifying a set of options in parentheses before the first variable name. For example

```
$HISTOGRAM (Y = 0,50    TAILS = 0,1) X $
```

would arrange that values between 0 and 50 would be included in the histogram as usual; values below 0 (the lower tail, given the code 0) would be omitted altogether, while those above 50 (the upper tail, given the code 1) would be given a single bar to themselves. More details are in the *User's Guide*, Section 5.4.1.

@@@@

1.9 Leaving Glim

When you have finished your analysis you will want to leave Glim and return to the operating system of your computer. To do this, type

```
$STOP
```

You can also end a *job*, i.e. go back to square one without leaving Glim, by typing

```
$END
```

(though this will not release the input channels—see Section 8.2—nor reset the random number seeds—see Section 3.5). This command may be useful if you want to start analysing a completely new set of data, or if your present analysis has got into such a tangle that all you can think of is to abandon it.

1.10 Significant figures

Glim normally displays all its output numbers with (at least) four significant figures. By typing

$ACCURACY n $

with n an integer between 1 and 9, this can be altered to a greater or smaller number. In Glim3 this will affect only the $LOOK (Section 1.6), $CALCU-LATE (Chapter 3) and $PRINT (Section 8.6) commands; in Glim77, all displayed output will be affected.

2
Simple linear models

2.1 Fitting a linear model

Now we know how to input data, let us use Glim to fit a multiple regression equation. Suppose that we have input the small data matrix from Table 1.3.1 and that we want to use it to obtain the regression of chest circumference on height and weight. We need to assume that the mean value of CHEST for given values of HEIGHT and WEIGHT is given by the *linear prediction* formula

$$E(CHEST) = \beta_0 + \beta_1 \times HEIGHT + \beta_2 \times WEIGHT$$

(where E(CHEST) is notation for the mean or expected value of CHEST) and that the CHEST values are distributed about their means with constant variance σ^2.

So far Glim knows only that we have three variables each with 10 values. We must tell the program which of our variables is to go on the left-hand side of the equation—which it is that is to be *predicted* or *explained* by the model. Glim calls this the *y-variable* (or *y*-variate), and we must issue a $YVARIABLE directive followed by the appropriate variable name. In our case we should type something like

`$YVARIABLE CHEST`

In the absence of any further indications, Glim assumes that ordinary multiple regression is what we want so we can go straight ahead with the fitting process. Our model contains the two *predictors* or *x-variables* HEIGHT and WEIGHT and to fit it we need to issue the command

`$FIT HEIGHT + WEIGHT $`

At last Glim does some work for us and produces some output. What you should see with Glim3 is three items something like

```
CYCLE        DEVIANCE         DF
   1          2.040           7
```

@@@@
The layout with Glim77 is slightly different—it would look like

```
deviance = 2.039713
    d.f. = 7
```

With Glim3 you will only get four significant figures under the DEVIANCE heading. With Glim77, you can get more (as shown here) by using the

16

$ACCURACY directive (see Section 1.10)
@@@@

For the time being you can ignore the CYCLE information—Glim77 will not give it anyway. For ordinary multiple regression iterative calculations are not needed and the CYCLE count will always be equal to 1. The other two quantities are actually quite familiar, consisting of the *residual sum of squares* about the regression and its degrees of freedom. The Glim term DEVIANCE always means something analogous to a residual sum of squares, though as we shall see it is a more general concept which arises in several different contexts.

The usual approach to multiple regression produces an analysis of variance with three sums of squares, those for regression and residual adding up to the total. This last sum of squares is in fact just the residual sum of squares from a linear model with nothing in it but a constant term

$E(CHEST) = \beta_0$

We can fit this by typing

$FIT $

and with Glim3 this produces the output

```
CYCLE      DEVIANCE        DF
  1          32.41          9
```

When we leave out the two predictor variables from the model the deviance increases from 2.040 to 32.41 so that the *extra deviance* is $32.41 - 2.04 = 30.37$ and this has $9 - 7 = 2$ d.f. This extra deviance is the same as the regression sum of squares so that we can now write down the analysis of variance table (Table 2.1.1). The divisions and subtractions are most

Table 2.1.1 Analysis of variance of chest data

	d.f.	SS	MS	F
Regression	2	30.37	15.18	52.34
Residual	7	2.04	0.29	
Total	9	32.41		

easily done on a calculator, though we shall see later that we can use Glim itself for the purpose.

@@@@
Glim77 will tell us the extra deviance directly as you will see if you type

$FIT : + HEIGHT + WEIGHT

Here the colon : is the *repetition symbol* and means 'repeat the last directive'. The second $FIT command (represented by the :) starts with a + sign and this means 'add the following *x*-variables to whatever is in the model at present'—in our case, to the constant term. The Glim77 output will look like

```
deviance = 32.40900
    d.f. =  9

deviance = 2.039713 (change =  -30.3693)
    d.f. = 7        (change =  -2      )
```

Note that we have typed two commands on the same line without waiting for the first one to be obeyed. This is convenient but makes it more difficult to match the output with the commands.
@@@@

$FIT commands starting with either a + or a − can be used in either Glim3 or Glim77 but the latter will show you the change in deviance and degrees of freedom. The repetition symbol : can also be used in both versions of Glim and we shall often encounter it. If the $FIT directive is followed only by an operator such as . or +, the previous model will be fitted over again.

2.2 Displaying the results

So far so good, but there are far more interesting aspects of the fitted model than the analysis of variance. To look at some of them we can use the $DISPLAY directive. We need to put one or more code letters following this directive to specify just what aspects of the fit we want to be displayed. For a start, let us look at the *estimates* of the parameters—this needs code letter E. Typing

```
$DISPLAY E $
```

produces

```
        estimate           s.e.    parameter
1        37.1514         6.65308    1
2       -0.153480      0.0375336    HEIG
3        0.267650      0.0275866    WEIG
scale parameter taken as  0.291388
```

The interpretation of this is quite straightforward once it is realized that the mysterious '1' under the parameter heading (or the even more mysterious %GM which Glim3 will give you[*]) is Glim notation for the intercept in the regression equation. The fitted model is

---

[*]The % sign is actually the *function symbol* mentioned by the ENVIRONMENT I directive. It may be different at your installation. Glim77 calls it the *system symbol*. GM stands for 'General Mean'; this will make more sense when we get to Chapter 4.

E(CHEST) = 37.15 − 0.1535 × HEIGHT + 0.2676 × WEIGHT

The standard errors of the regression coefficients are 0.0375 and 0.0276 (Glim3 may print 0.3753E-01 and 0.2759E-01, using 'scientific notation' with E meaning "times 10 to the power). From these we can calculate t-values of $0.1534/0.0375 = 4.09$ and $0.2676/0.0276 = 9.69$, both highly significant on 7 d.f.

How well does our model fit the data? To some extent this is measured by the residual mean square of 0.29. A more accurate version of this is the *scale parameter* equal to 0.291388 which appears at the bottom of the displayed coefficients. The square root of this is 0.540 so that HEIGHT and WEIGHT are predicting chest circumference within a range of roughly ± 1 cm. To go further than this we should look at the predictions from the model and compare them with the actual observations. To do this, use the $DISPLAY directive with code letter R—type

$DISPLAY R $

The letter R stands for *residuals*. This produces the table

unit	observed	fitted	residual
1	30.0000	30.5994	−0.59935
2	29.4000	29.0423	0.35773
3	26.1000	26.1384	−0.03838
4	25.7000	25.8464	−0.14639
5	27.9000	28.1974	−0.29741
6	31.7000	31.4173	0.28272
7	27.4000	27.8448	−0.44478
8	29.0000	28.9154	0.08462
9	30.3000	30.5456	−0.24562
10	29.4000	28.3531	1.04686

The 'observed' column contains the data values for chest circumference, the y-variable, while the 'fitted' column has the values worked out from the regression equation—for example the first subject has a height of 167.9 and a weight of 71.8 so that his fitted value is

37.15 − 0.1536 × 167.9 + 0.2676 × 71.8

The 'residual' column contains the differences (observed − fitted). These do not show any very obvious pattern, though the last one may be a bit big. We shall discuss examination of the residuals further in Chapter 4.

2.3 A larger problem

Table 2.3.1 contains some data relating to the number of bird species nesting on islands round the coast of the British Isles. There are seven variables and 43 units; we shall call the y-variable SPEC and the six x-variables are

DIST   distance from the mainland (km)
LAT   N latitude (degrees)
LONG   W longitude (degrees)
AREA   area (ha)
HAB   number of habitats
ELEV   maximum elevation (m)

Suppose we type these data into a disk file and contrive that this file is attached to channel 8. Then we can read in the data by typing

```
$UNITS 43
$DATA SPEC DIST LAT LONG AREA HAB ELEV
$DINPUT 8 $
```

To fit the model and display the estimated coefficients, type

```
$YVARIABLE SPEC
$FIT DIST + LAT + LONG + AREA + HAB + ELEV
$DISPLAY E $
```

Table 2.3.1 Bird species dataset

SPEC	DIST	LAT	LONG	AREA	HAB	ELEV
4	49.9	49.54	6.22	21.4	8	18.0
45	3.2	52.46	4.48	179.8	19	167.0
7	5.2	56.05	2.38	2.8	8	82.0
36	35.4	54.03	4.50	249.5	18	127.9
51	15.7	56.10	6.46	7418.1	22	103.3
17	35.4	58.54	2.39	85.1	16	51.2
2	4.8	56.03	3.12	6.5	7	5.1
2	4.8	51.24	2.51	0.8	3	3.7
19	0.8	58.32	4.20	187.9	14	75.2
26	249.4	60.37	0.50	4011.5	20	158.4
8	6.1	51.22	3.08	21.1	7	18.3
2	54.7	57.59	7.24	19.0	9	32.0
5	3.5	55.17	5.37	5.7	4	27.1
6	44.5	49.54	6.19	38.6	11	24.7
3	64.4	57.42	7.36	13.8	6	37.5
10	0.4	52.05	− 1.80	108.5	11	7.6
6	1.1	55.39	4.56	6.9	6	3.1
8	3.4	56.00	3.13	1.4	7	14.6
35	0.2	55.41	2.47	541.9	17	21.0
4	0.4	51.26	5.14	1.8	5	33.5
41	8.1	56.50	6.10	642.3	21	137.4
13	225.3	60.24	0.44	265.3	11	51.8

14	0.8	53.19	4.02	28.3	6	49.7
27	0.8	51.53	5.21	243.0	19	135.9
29	45.1	49.54	6.20	109.4	16	24.1
2	1.0	51.36	4.44	5.6	7	39.0
31	2.6	55.16	5.34	18.6	10	123.4
24	0.4	52.58	−0.19	334.1	9	12.2
42	1.2	51.45	5.18	292.4	18	78.9
4	49.9	59.04	4.24	13.8	3	12.2
53	25.7	56.31	6.51	7652.9	24	140.1
2	15.7	53.55	10.17	0.8	4	4.9
28	4.8	52.05	10.30	631.	16	292.8
40	6.4	51.26	9.37	639.1	20	133.4
25	2.8	54.41	5.32	32.0	11	33.2
9	3.2	53.34	10.07	639.1	7	62.8
6	11.3	51.48	10.33	17.8	8	13.7
1	0.8	54.48	8.34	4.9	4	17.4
33	4.2	53.32	10.18	961.5	12	89.0
2	3.2	54.34	8.33	1.2	3	13.7
16	3.2	54.06	10.09	360.4	7	70.1
16	6.4	54.28	8.40	84.6	11	26.5
19	11.3	55.16	7.12	331.4	12	82.3

This produces

```
deviance = 1481.653
    d.f. =    36

          estimate              s.e.       parameter
   1      -8.31604          23.5192        1
   2      -0.0610732         0.0233147     DIST
   3       0.0592802         0.422211      LAT
   4      -0.0414930         0.375510      LONG
   5       0.00118933        0.000760019   AREA
   6       2.01149           0.273550      HAB
   7       0.0276912         0.0241715     ELEV
   scale parameter taken as   41.1570
```

It is worth noting that the intercept value of −8.316 with the model in its present form is the expected number of species on an island of zero size, touching the mainland and situated on the equator. It is perhaps not too disturbing that this nonsense quantity comes out negative.

The R display for this example is shown in Table 2.3.2 and you should examine it, but there are several other ways in which we can study the goodness of fit. Every time Glim obeys a $FIT command it not only calculates the fitted values (those given in the R display) but also stores them in a new vector with the name %FV. This is a *system vector*; the *system symbol* % as the first character of its name distinguishes it from names that you provide yourself, but we can use the vector just like any other. For instance we can plot the observed against the fitted values—

Table 2.3.2 Bird species dataset—residuals from 6-variable regression

unit	observed	fitted	residual
1	4.000	7.931	− 3.931
2	45.000	37.469	7.531
3	7.000	12.956	− 5.956
4	36.000	32.583	3.417
5	51.000	49.719	1.281
6	17.000	26.596	− 9.596
7	2.000	8.812	− 6.812
8	2.000	0.462	1.538
9	19.000	25.385	− 6.385
10	26.000	29.397	− 3.397
11	8.000	8.832	− 0.832
12	2.000	10.469	− 8.469
13	5.000	3.321	1.679
14	6.000	14.502	− 8.502
15	3.000	3.973	− 0.973
16	10.000	17.286	− 7.286
17	6.000	6.874	− 0.874
18	8.000	9.153	− 1.153
19	35.000	30.275	4.725
20	4.000	5.472	− 1.472
21	41.000	41.096	− 0.096
22	13.000	5.353	7.647
23	14.000	8.100	5.900
24	27.000	36.744	− 9.744
25	29.000	24.590	4.410
26	2.000	9.650	− 7.650
27	31.000	18.128	12.872
28	24.000	13.623	10.377
29	42.000	33.185	8.815
30	4.000	− 1.651	5.651
31	53.000	54.440	− 1.440
32	2.000	1.660	0.340
33	28.000	35.091	− 7.091
34	40.000	38.627	1.373
35	25.000	17.601	7.399
36	9.000	10.812	− 1.812
37	6.000	10.109	− 4.109

38	1.000	3.052	− 2.052
39	33.000	21.912	11.088
40	2.000	0.779	1.221
41	16.000	10.725	5.275
42	16.000	17.123	− 1.123
43	19.000	20.779	− 1.779

$PLOT SPEC %FV $

with the result in Fig. 2.3.1. The line of equality has been drawn in by hand and there do not seem to be systematic departures from it, but, as the R display will have shown you, the model has given one negative prediction which does not make a lot of sense. We shall see in the next chapter how to make a plot similar to this which is somewhat easier to read by eye.

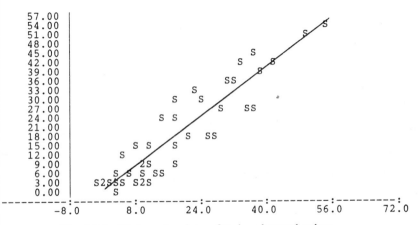

Fig. 2.3.1   Bird species data—fitted v observed values.

Going back to the E display, we may notice that the coefficients of LAT and LONG are very small compared to their standard errors, suggesting that the 'true' coefficients of these two x-variables might be zero. Type

$FIT − LAT − LONG $DISPLAY E $

to find the effect of taking them out of the model. The result is

```
deviance = 1483.240 (change =  +1.58719)
    d.f. =     38   (change =  +2       )

            estimate                s.e.      parameter
    1       -5.38295             2.24407      1
    2       -0.0587212           0.0194179    DIST
    3        0.00120048          0.000711583  AREA
    4        2.01560             0.252471     HAB
    5        0.0268414           0.0219607    ELEV
    scale parameter taken as   39.0326
```

Removing the two x-variables can only increase the residual sum of squares, and we can see that the increase is actually $+1.587$. This is associated with two extra degrees of freedom, and the increase per d.f. of $1.587/2 = 0.794$ is much smaller than the scale parameter of 39.03 which is what we might expect by chance. We can set out a formal analysis of variance table which tests the hypothesis that both of the true coefficients of LAT and LONG are equal to 0 (Table 2.3.3), where the total sum of squares

Table 2.3.3   Analysis of variance for bird species data

	d.f.	SS	MS
Regression on 4 x-variables	4	8363.737	
Extra for including LAT and LONG	2	1.587	0.794
Residual	36	1481.653	41.157
Total	42	9846.977	

can be found as the deviance from the command

`$FIT`

As above, the mean square for the combined effects of LAT and LONG is much smaller than the residual mean square—suspiciously so, indeed, for the value of F^2_{36} of $0.794/41.157 = 0.019$ gives $P = 0.98$. Can there be something wrong with our model? We shall see.

2.4 Regression with replicates

Consider the data in Table 2.4.1. These represent the responses of a measuring device to five concentrations of a chemical substance; the problem is to describe how the response varies with concentration. You will see that there are three replicate readings at each concentration. So far in our examples we have used the deviations from the values predicted by the fitted model to assess the variability of the y-variable, but here we can get a direct estimate from the differences between the replicates, which for any given value of x all have the same expectation.

Let us do this part of the calculations first by constructing a one-way analysis of variance. For this purpose we need to fit a model of the form

$$E(y_{ij}) = \mu + \alpha_i$$

Here y_{ij} is the jth reading in the ith group, μ is a parameter common to all the readings and α_i is a parameter common to all the readings in the ith group. We assume that the ys vary about their means with constant variance σ^2.

This looks like a linear model—there are no powers or products of the

Table 2.4.1 Response (3 replicates) v concentration in a chemical measurement

Concentration	Response		
1.0	31.5	34.3	34.0
3.0	73.0	85.5	75.6
6.0	140.2	144.5	151.0
10.0	200.8	195.1	191.2
15.0	222.9	212.9	228.8

parameters in it—but it seems to lack any x-variables. Let us invent some. Suppose we have a set of five x-variables such that x_i ($i = 1, ..., 5$) takes the value 1 for all readings in group i and the value 0 everywhere else. Now fit a linear model

$$E(y_{ij}) = \beta_0 + \beta_1 x_1 + \cdots + \beta_5 x_5$$

Consider a reading in (say) group 3. For this reading, x_3 is equal to 1 and all the other xs are equal to 0, so the expected value is

$$E(y_{3j}) = \beta_0 + \beta_3$$

It can be seen that we have exactly the same model as before with no more than a change of notation; μ has become β_0 and α_i has become β_i. We can go ahead and fit the model using Glim.

What a nuisance it is going to be, though, to have to type in 15 values of each of five different x-variables. Let us make Glim do this for us. Remember that at this stage we are treating the five concentrations just as five independent groups of readings and taking no account of the actual concentrations themselves. Let us label the groups from 1 to 5 and type in these group labels along with the rest of the data as follows:

```
$UNITS 15      $DATA GRP X Y      $READ
1   1.0   31.5      1   1.0   34.3      1   1.0   34.0
2   3.0   73.0      2   3.0   85.5      2   3.0   75.6
3   6.0  140.2      3   6.0  144.5      3   6.0  151.0
4  10.0  200.8      4  10.0  195.1      4  10.0  191.2
5  15.0  222.9      5  15.0  212.9      5  15.0  228.8
```

The labels are in the vector GRP (note that we must repeat the labels and the x-values for each of the 15 readings). We now need to tell Glim that the contents of GRP are not continuous quantities but should be treated as ordinal numbers, 1st, 2nd, 3rd, ... Glim calls a vector like this a *factor*, in our case with 5 *levels* (we met factors before in Sections 1.7 and 1.8). To convey this information type

```
$FACTOR GRP 5 $
```

Behind the scenes Glim will now construct the necessary x-variables so that if you now type

```
$YVARIABLE Y      $FIT GRP      $DISPLAY E $
```

you will get

```
deviance = 326.6866
    d.f. =   10
            estimate              s.e.      parameter
    1       33.2667            3.29993      1
    2       44.7667            4.66681      GRP(2)
    3       111.967            4.66681      GRP(3)
    4       162.433            4.66681      GRP(4)
    5       188.267            4.66681      GRP(5)
    scale parameter taken as   32.6687
```

This takes a little explanation. Rows 2 to 5 presumably relate to groups 2 to 5, but what has happened to group 1? To see what is going on we need to look back at the model. Suppose we increase β_0 by an arbitrary amount—say 10—and at the same time reduce each of β_1 to β_5 by the same amount. You will see that this makes no difference at all to the expectations. The model as we have written it does not provide unique estimates of its parameters, and we can in fact fix one of the estimates at an arbitrary value provided we adjust all the others to fit. Glim chooses to set the estimate belonging to the first level of the factor to zero, so that you can imagine a row for GRP(1) with an estimate of 0.00 and a standard error of 0.00 as well. This means that the estimate of 44.77 for GRP(2) actually estimates the difference GRP(2)—GRP(1), and the standard error is the standard error of this difference. Note carefully that this difference does *not* depend upon Glim's arbitrary setting of GRP(1); if Glim had set GRP(1) = 100 it would have found GRP(2) = 144.77, with the intercept at -66.73 to compensate. The expected values for the 5 groups are in fact

$$E(y_{1j}) = 33.2667 + 0 \qquad = \quad 32.27$$
$$E(y_{2j}) = 33.2667 + 44.7667 \qquad = \quad 44.03$$
$$E(y_{3j}) = 33.2667 + 111.967 \qquad = \quad 145.23$$
$$E(y_{4j}) = 33.2667 + 162.433 \qquad = \quad 195.70$$
$$E(y_{5j}) = 33.2667 + 188.267 \qquad = \quad 221.53$$

(You can verify these by typing $DISPLAY R.) These are just the observed means of the five groups.

All this is slightly by the way for our example; more to the point is the estimate of the residual variance given by the scale parameter of 32.6687. We would like to use this from now on as the basis for the standard errors and F-values, and we can arrange this by typing

```
$SCALE 32.6687    $
```

If we now repeat the $FIT GRP command we get

```
scaled deviance =  9.99999
         d.f. = 10
```

and we see that the residual mean square is now 1.0000 (to within rounding errors) in terms of the new scale parameter.

Now let us see how well a straight-line model

$E(y) = \beta_0 + \beta_1 x$

fits the data. Typing

```
$FIT X     $DISPLAY E
```

produces

```
scaled deviance = 181.7682
         d.f. =  13

         estimate        s.e.      parameter
   1       40.1756      2.53234     1
   2       13.5111      0.293982    X
   scale parameter taken as   32.6687
```

The change in scaled deviance in going from the GRP model to the X model is 171.7682 with 3 d.f. This is just the scaled sum of squares of the deviations of the group means from the straight line (remember that the GRP model fits the group means exactly). The mean extra deviance is $171.7682/3 = 57.26$ and this is an F-ratio with 3 and 10 d.f. This is enormously significant, indicating that the group means are farther from the straight line than the within-group variability would lead us to expect. To see what is going on, plot y against x by typing

```
$PLOT Y X$
```

to get Fig. 2.4.1 (we should really have done this before starting on any analysis). It is clear that the relation is curved, not straight.

Perhaps a better model would include a squared term. Let us see. Type

```
$DATA X2    $READ
1 1 1 9 9 9 36 36 36 100 100 100 225
$FIT +X2    $DISPLAY E $
```

This produces

```
scaled deviance = 12.27718 (change =  -169.491)
         d.f. = 12        (change =    -1     )
```

```
              estimate              s.e.        parameter
    1         3.78333             3.77183       1
    2        28.8260             1.21254        X
    3        -0.954342           0.0733045      X2
    scale parameter taken as   32.6687
```

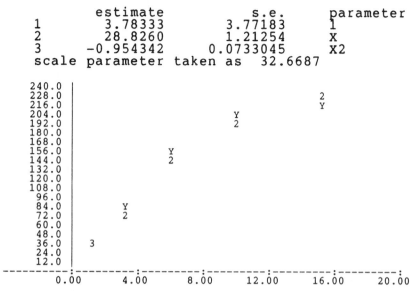

Fig. 2.4.1   Plot of y against x from data in Table 2.4.1

Notice first that the coefficient of x^2 is highly significant with a t-value of 13.02 on 10 d.f. (we are still using the scale parameter to estimate the standard errors). There is no doubt that including this term in the model has improved the fit. Secondly, the extra deviance for deviations from the line is now $12.27718 - 10 = 2.27718$ on 2 d.f. The mean extra deviance is 1.14, an F-ratio with 2 and 10 d.f. and a very satisfactory level.

We can take the analysis a stage further if we notice that the intercept of 3.783 scarcely exceeds its standard error. What about a quadratic passing through the origin as a possible model? To try this we must remove the intercept from the model by typing

$FIT -%GM $

with Glim3 or

$FIT -1 $

with Glim77. Followed by $DISPLAY E this produces

```
scaled deviance = 13.28329 (change =  +1.00611)
          d.f. = 13        (change =  +1       )

              estimate              s.e.        parameter
    1        29.8614             0.636238       X
    2        -1.00883            0.0492153      X2
    scale parameter taken as   32.6687
```

The extra deviance for deviations from the curve is now 3.28329 giving an F-ratio of 1.09 with 3 and 10 d.f., again entirely acceptable. It seems that a response curve

$$y = 29.861x - 1.009x^2$$

with a standard deviation about the curve of $\sqrt{32.6687} = 5.72$ provides a reasonable description of the data.

2.5 A cautionary example

Table 2.5.1 contains a rather simple body of data. The y-values are the logarithms of daily catches in an insect trap and the x-variables are minimum and maximum temperature. Clearly the catch increases with temperature in a general way and we would like to investigate this.

Put the data into a file, connect the file to channel 8 (say) and type

```
$UNITS 20
$DATA Y MIN MAX
$DINPUT 8
$YVARIABLE Y
$FIT $DIS E $FIT MIN $DIS E $FIT MAX $DIS E $FIT MIN + MAX $DIS E
```

Table 2.5.1 Catches in insect trap v temperature

	Temperature	
Log catch	Min	Max
1.23	6	15
1.30	8	18
1.18	6	17
1.30	4	12
1.26	3	12
1.49	3	13
1.48	5	18
1.54	6	17
1.51	8	18
1.70	8	20
1.82	10	21
1.68	12	22
1.75	13	21
1.91	13	20
1.34	10	21
1.70	8	20
1.66	7	15
1.41	7	13
1.48	5	12
1.53	8	14

The result of this is

```
deviance  =   0.81745
   d.f.  = 19

             estimate           s.e.        parameter
    1            1.514        0.04638        1
    scale parameter taken as  0.04302

deviance  =   0.46473
   d.f.  = 18

             estimate           s.e.        parameter
    1            1.169        0.09994        1
    2          0.04596        0.01243        MIN
    scale parameter taken as  0.02582

deviance  =   0.56272
   d.f.  = 18

             estimate           s.e.        parameter
    1           0.9456        0.2028         1
    2          0.03350        0.01174        MAX
    scale parameter taken as  0.03126

deviance  =   0.46303
   d.f.  = 17

             estimate           s.e.        parameter
    1            1.123        0.2108         1
    2          0.04157        0.02173        MIN
    3         0.004656        0.01864        MAX
    scale parameter taken as  0.02724
```

(Notice that we have again put all the $FIT and $DISPLAY commands on one line. This is handy for typing but a bit confusing when it comes to matching up the output with the commands.)

We might look first at the two-variable regression. The t-values for MIN and MAX are $0.04157/0.2173 = 1.91$ and $0.004656/0.01864 = 0.25$, neither of them reaching the magic 5% level of significance. Are we to deduce that neither of the two temperature measurements is affecting the level of catch in the trap? Surely not; the t-values from the single-variable regressions are $0.04596/0.01243 = 3.70$ for MIN and $0.03350/0.01174 = 2.85$ for MAX, both highly significant. What is going on here?

To make things clearer, look again at the problem, this time in terms of the residual deviances. These are given in Table 2.5.2 and from these we can build an analysis of variance (Table 2.5.3).

Now the position should be clearer. The regression on either MIN or MAX taken alone is highly significant. The extra contribution due to adding in the second x-variable, however, is much less impressive, no matter which one we start with. The data are trying to tell us that one or other of the x-variables is definitely required to describe the behaviour of y, but that there is no real

Table 2.5.2 Insect trap data: residual deviances

	d.f.	deviance
Constant term only	19	0.81745
Regression on MIN	18	0.46473
Regression on MAX	18	0.56272
Regression on MIN and MAX	17	0.46303

Table 2.5.3 Insect trap data: analysis of variance

	d.f.	SS	MS	F
Regression on MIN alone	1	0.35272	0.3527	12.95
extra for including MAX	1	0.00170	0.0017	0.06
Regression on MAX alone	1	0.25472	0.2547	9.35
extra for including MIN	1	0.09970	0.0997	3.66
Regression on MIN and MAX	2	0.35442	0.1772	6.51
Residual	17	0.46303	0.0272	
Total	19	0.81745		

need for both. Roughly speaking, it is temperature *per se* that is affecting the catch, and we can measure temperature by MIN or MAX indifferently.

The basic point is that the *t*-tests of coefficients in a multiple regression (and the *F*-tests for groups of coefficients) always test the significance of the corresponding *x*-variables *after all the others have been fitted*. The 'significance' of an *x*-variable can be quite different according to which of the other *x*s are in the model being considered. This apparent paradox becomes straightforward when we think in modelling terms. It is quite natural that a particular *x* may be a very important predictor on its own, but that its importance is much reduced once another *x* has been allowed for. This is always liable to happen when (as in our example) the two or more *x*s in question are fairly highly correlated so that they all tell more or less the same story. Such *x*s are said to be *confounded* with each other.

3
Calculations

3.1 Arithmetic

Very often we need to do some arithmetic on our data before fitting a model. We may for instance need to transform one or more of the variables, or perhaps to combine two variables together to form a third. The Glim $CALCULATE directive enables us to do this very conveniently. If Y and Z are the names of two variables containing data we can issue commands such as

```
$CALCULATE X = Y + Z
$CALCULATE W = X - Z
```

Each item of the variable on the left is replaced by the value calculated from the corresponding items from the variables on the right. If the variables X and W in our example already have data in them, these are over-written by the new values; otherwise new variables are formed with the same number of items as the existing ones. As a tiny example, the directives above cause

Y	Z		X	W
1	2		3	1
3	-4	to produce	-1	-3
7	0		7	5

The expressions on the right-hand side of the equals sign can be almost arbitrarily complicated. They are written according to the rules of Fortran (more or less—see Appendix B) which differ only slightly from those of computer languages such as Basic and Pascal. Multiplication is denoted by $*$ and exponentiation (raising to a power) by $**$ so that $X**2$ calls for each item of X to be squared (users of other languages may be accustomed to $\wedge$ in this context). Thus to form the so-called ponderal index, height/(weight)$^{1/3}$, the command might be

```
$CALCULATE POND = HEIG/WEIG**(1/3)     $
```

Fortran programmers should note that all Glim quantities are reals, not integers, so that $1/3 = 0.333333...$ as required.

Of course $CALCULATE can always be abbreviated to $CAL in the usual way. With several $CALCULATEs in sequence, we can use the : repetition symbol to save typing. The first two commands above could be written

```
$CALCULATE X = Y + Z    :W = X - Z    $
```

32

If there is no equals sign following a $CALCULATE directive, the value of the expression is simply displayed on the screen—thus

$CALCULATE 2 + 2 $

displays 4.000. This provides a simple calculator facility.

3.2 Scalars

As well as variables in vectors, Glim allows for the use of single numbers which it calls *scalars*. These too have names, but the names are laid down in advance and are distinguished from names made up by the user by beginning with the *system symbol* (or *function symbol* as Glim3 calls it), usually %. There are 26 scalars available for general use, with names %A, %B, ..., %Z. These can be used in $CALCULATE commands—for example

$CALCULATE %A = 2   : X = X - %A $

will subtract 2 from each of the values in X.

As well as these, there are several *system scalars* which are given values by Glim itself as a result of its calculations. A list is given in Section 10.4. As examples I can mention %NU which contains the number of units set by the most recent $UNITS directive; %DV and %DF which contain the deviance and degrees of freedom from the most recent $FIT; and %PI which is initially set to 3.14159...

Some slightly odd situations arise when the left-hand side of a $CALCULATE command is a scalar while the right-hand side includes a vector. The arithmetic will be done for each item in the vector successively. Thus

$CALCULATE %A = X $

will set %A to the value of the last item in X, and

$CALCULATE %A = 1   : %A = %A * X $

sets %A to the product of all the items in X.

3.3 Functions

Built into Glim is a small repertoire of arithmetic functions which can be used within expressions in a $CALCULATE directive. These are

%ANG(X)	value is	$\sin^{-1}(\sqrt{X})$ in radians. X must lie between 0 and 1.
%EXP(X)		e^X
%LOG(X)		$\log_e(X)$
%SIN(X)		sin (X). X must be in radians.
%SQR(X)		$\sqrt{X}$
%NP(X)		Normal probability integral from $-\infty$ to X

| %ND(X) | Normal equivalent Deviate of X (the inverse of %NP). X must lie between 0 and 1. |
| %TR(X) | Integral part of X, Truncating towards zero. |

Attempts to do impossible operations, such as taking the logarithm of zero or the square root of a negative quantity, are not fatal but give a result of 0 along with a warning message on the terminal.

The function %CU(X) produces a vector of cumulative sums. Thus if

X contains	1	$CALCULATE Y = %CU(X) $ gives Y as	1
	3		4
	2		6
	7		13

The command

```
$CALCULATE %A = %CU(X)    $
```

puts the sum of the values in X into %A. The vector %CU(1) is treated as if its values were 1, 2, ... up to the current number of units.

The function %GL(n1,n2) where n1 and n2 are integers or scalars with integer values means 'Generate the numbers from 1 to n1 in blocks of size n2'. GL stands for 'Generate Levels' and the %GL function is intended for the input of factor levels when the data are arranged systematically. As an example

```
$CALCULATE X = %GL(3,2)    $
```

produces the numbers 1 1 2 2 3 3 1 1 2 2 3 3 1 1 2 ...in X, repeating until the current number of units has been generated. In Section 2.4 we could have used

```
$CALCULATE GRP = %GL(5,3)    $
```

to fill in the group codes instead of typing them in explicitly.

3.4 Relations and conditionals

Glim has a set of functions for comparing two quantities. These are

%LT(X,Y)	true if $X < Y$
%LE(X,Y)	true if $X \leqslant Y$
%EQ(X,Y)	true if $X = Y$
%NE(X,Y)	true if $X \neq Y$
%GE(X,Y)	true if $X \geqslant Y$

%GT(X,Y)   true if X > Y

The value of each function is 1 if the result is true and 0 if it is false. If one of these functions has a vector argument it will produce a vector of 0s and 1s as its result, so that

```
$CALCULATE EQL = %EQ(X,Y)    $
```

will produce a variable EQL which is 1 when corresponding values of X and Y are equal and 0 when they are not. It is quite easy to play simple tricks with these functions. For example, we can get a variable whose items are equal to the smaller of those in X and Y by typing

```
$CALCULATE SMALL = X * %LE(X,Y) + Y * %GT(X,Y)    $
```

There is also a conditional (IF) construction in Glim, though it takes a bit of getting used to. The value of the function

%IF (relation, expression, expression)

is equal to the first expression if the relation is true (i.e. has a non-zero value), the second expression if the relation is false (has a zero value). We can for instance replace the values in a vector X by their absolute values by typing

```
$CALCULATE X = %IF(%LT(X,0),-X,X)    $
```

which can be read as 'if X is less than 0 then −X, else X'. Ingenious programmers will perceive that extremely complex conditions can be constructed, though they may not be very easy to read.

@@@@
Glim77 allows for more readable versions of the comparison functions, using the symbols =, <, > and / (meaning 'not'). Thus X > = Y means the same as %GE(X,Y). The symbol for %EQ is ==.

Glim77 also contains operators for 'logical' quantities whose values are zero (meaning 'false') or non-zero (meaning 'true'). These are & (and), ? (or), and / (not). We have that

X & Y	means the same as	(X/=0) * (Y/=0)
X ? Y	means the same as	1−(X==0) * (Y==0)
/X	means the same as	(X==0)

These operators are governed by precedence rules like the ordinary arithmetic operators, but it is not necessary to remember these if you are fairly lavish with the use of brackets. As a horrid example,

%EQ(2,2) + %EQ(3,3) is equal to 2

but

2 == 2 + 3 == 3 is equal to 0

since it is interpreted as $(2 == (2+3) == 3)$.
@@@@

It should be noted that both the expressions in an %IF function are always evaluated. You cannot evade a warning message by typing

```
$CALCULATE Z = %IF(%NE(X,0),1/X,0)    $
```

though you need not have bothered since 'impossible' values like $1/0$ are set to zero anyway.

3.5 Random numbers

A function %SR(n) produces a pseudo-random number every time it is referred to (SR stands for Standard Random). The argument n should be an integer. If $n = 0$ the random numbers will be uniformly distributed in the range 0–1. If n is greater than 0 the random numbers will be uniformly distributed integers in the range 0 to n. The command

```
$CALCULATE X = %SR(0)    $
```

will fill the vector X with random numbers between 0 and 1. If Normally distributed random numbers are needed, the %ND function can be used. Thus to get random Normal variates with mean 50 and standard deviation 10, type

```
$CALCULATE X = 50 + 10 * %ND(%SR(0))    $
```

The random number generator that Glim uses has to be initialized by setting three integer *seed* values—the first two of these must be less than 4096, the third less than 2048. Glim does this for you at start-up time using values of its own (you can find out what these are by the command $ENVIRONMENT R) but you can set them yourself at any time by a command such as

```
$SSEED 2571 3024 97 $
```

One, two or three seeds can be set by this command; Glim will supply values for any that you leave out.

For those interested in such things, the standard random number generator used in the %SR function is

$$x_{i+1} = 8\ 404\ 997 x_1 + 1 \qquad \mathrm{mod}\ 2^{35}$$

This produces 2^{35} different numbers before repeating.

The piece of the Glim program which calculates the random numbers is written in standard Fortran and should give the same results on any computer (subject to some possible limitations due to word-length). For this reason it may be a bit slow. Installations can install their own random number generator and seeding routines, specially coded to give high speed or some other desirable property. When these are present they can be accessed by the function %LR (for Local Random) and the command $LSEED, which should behave just like the standard versions. Those interested should consult the local experts for details.

3.6 Sorting

This is a convenient place to describe the Glim facility for sorting data. Like several other Glim directives, the $SORT command is powerful but can get a bit complicated. First of all, if X is a vector the command

$SORT X $

will replace the values in X by the same values sorted into ascending order. If before the command X contained the values 1,3,2,4,0 then after the command it would contain 0,1,2,3,4.

If Y is another vector. the command

$SORT Y X $

puts the sorted values from X into Y. You have to remember that the destination vector comes first and the source vector second.

More complicated is the three-vector command

$SORT Y X Z $

This uses the third vector Z as a key and goes through the motions of sorting it into ascending order. The items in Z are not actually moved, but those in X are re-ordered at the same time and the results are stored in Y. Thus the first value in Y is the value in X which corresponds to the smallest value in Z; the next value in Y is the value in X corresponding to the second smallest value in Z; and so on. You can see how this works by showing how the command applied to

X	Z		Y
7	2	produces	11
9	3		7
11	0		9
13	6		13

You might also like to show that the commands

$CALCULATE Z = −X
$SORT Z X Z $

sort the values in X into descending sequence and put them into Z.

Scalar values can also be used for the second or third argument of the $SORT command. In this context an integer k should be thought of as standing for a vector with items k, $k + 1$, $k + 2$,...,n,1,2,... where n stands for the number of items in the vector. The value $k = 1$ is the most often useful. For example

```
$SORT Y 1 X $
```

means 'sort the items 1,2,3,... according to the values in X and put the results into Y'. Y will now contain the positions of the ranked items in X, with the first item of Y giving the position of the smallest item in X, the next item in Y the position of the second smallest item in X, and so on. As an exercise, show how you can use two $SORT commands with 1 as the second argument to get the actual ranks of the items in X into Y.

As another application we can generate a random permutation of the integers 1 to n by the commands

```
$CALCULATE W = %SR(0)    !GENERATE N RANDOM NUMBERS IN W
$SORT PERM 1 W    $      !SORT 1 TO N INTO RANDOM SEQUENCE IN PERM
```

The command

```
$SORT Y X 2 $
```

produces in Y the items in X shifted one place along, with the first item in X put into the second item in Y, the second item in X into the third item of Y, and so on. Note however that the last item in X goes into the first item in Y which may or may not be what you want to happen.

3.7 Normal plotting

As an example of data manipulation, let us take the residuals from the fit of the simple model in Section 2.1 and plot them against their expected positions in a sample of the same size from a Normal distribution with mean 0 and standard deviation 1. This technique of forming a *Normal plot* is very useful for getting an idea of distributional shape and especially for examining visually whether apparently extreme values in a sample are really excessively large or small.

As we noted earlier, after every $FIT command Glim stores the fitted values in the system vector %FV. We can use this to calculate the residuals in our example simply by typing

```
$CALCULATE RESID = CHEST - %FV $
```

Next we sort these into ascending order:

```
$SORT RESID $
```

We can approximate the expected positions in a Normal sample by calculating the fractions $\frac{1}{2}/10$, $1\frac{1}{2}/10$, $2\frac{1}{2}/10$, ..., $9\frac{1}{2}/10$ and obtaining the Normal equivalent deviates that correspond to these. Type

```
$CALCULATE POS = %CU(1)    !GENERATE THE NUMBERS 1 TO 10
```

```
: POS = (POS - 0.5) / 10      !GENERATE THE FRACTIONS
: POS = %ND(POS)              !AND THEIR NORMAL DEVIATES
```

Finally we need

`$PLOT POS RESID $`

with the result shown in Fig. 3.7.1. It really does look as if one of the residuals is rather large compared to the others.

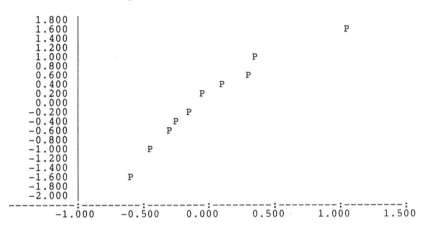

Fig. 3.7.1   Normal plot of residuals from data in Table 1.3.1

3.8 Suffices

Single items in a vector can be referred to by using suffices, just as in most ordinary programming languages. For example, X(5) means the fifth item in the vector X—it is of course a scalar quantity. We could write instructions such as

```
$CALCULATE X(5) = 10.2
: %C = X(5)
: Y = X(5) $
```

The last instruction copies X(5) to all the items of Y.

As usual, the Glim suffix facility allows you to play some clever tricks. Suppose S is a vector with three integer items. Then the command

```
$CALCULATE X(S) = X(S) + 2     $
```

will add 2 only to the three items of X which are pointed to by the items of S. Of course the vector S will not have the standard length set by the current $UNITS directive. If its contents are to be input by $DATA and $READ directives, it must first be *declared* by a directive

`$VARIABLE 3 S $`

@@@@
In Glim77, you have a more convenient way of specifying the contents of S
using the $ASSIGN directive (Section 1.3).
@@@@

Suffices normally run from 1 upwards, but the $CALCULATE directive
gives a special meaning to a suffix with the value zero. If an item with a zero
suffix turns up on the right-hand side of the equals sign, the value is taken to
be zero. If such an item appears on the left-hand side of the equals sign, no
value from the right-hand side is stored. Suppose we have a vector K of 0s
and 1s (maybe the result of a comparison operation) and we wish to extract
from the vector X the values corresponding to the 1s in K and put them into a
vector Y. The first thing to do is to calculate the length of Y and to declare it
to have this length:

```
$CALCULATE %L = %CU(K)        !NUMBER OF 1'S IN K TO %L
$VARIABLE %L Y $
```

Now we replace the 1s in K by 1, 2, 3, ... and use these to store the required
values in Y:

```
$CALCULATE J = K * %CU(K)
: Y(J) = X $
```

Since expressions can be used in suffices, we could combine the last two
commands (and do without the vector J) by typing

```
$CALCULATE Y(K * %CU(K)) = X $
```

As an exercise, show how you can overwrite the positions specified by the
1s in K in an existing vector Z with the values from Y without affecting the
other values in Z.

3.9 More on the Bird species dataset

Let us return to the Bird species dataset that we started to analyse in Chapter
2. We can calculate the residuals following the fit of the 6 x-variables by
typing

```
$CALCULATE RES = SPEC - %FV
```

and if we now plot these against the fitted values in %FV we get the plot in
Fig. 3.9.1. There is some hint that the variability of the residuals increases
with %FV, at least up to around %FV = 20; this is not surprising since it is
unlikely that the precision of a predicted 2 species will be the same as that of
a predicted 20 species. Also you will remember that our model gave a
negative prediction and an unpleasantly small F-ratio on removing the x-
variables LAT and LONG.

Suppose we try the effect of transforming the number of species to
logarithms. This will be appropriate if the effect of increasing an x-variable is

to increase the number of species by a constant factor rather than by a constant amount and if the

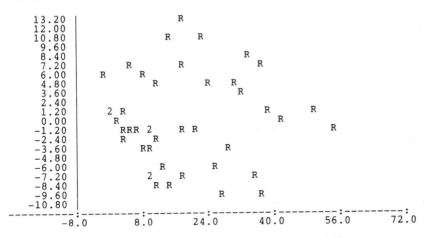

Fig. 3.9.1   Residuals v fitted values, SPEC as y-variable

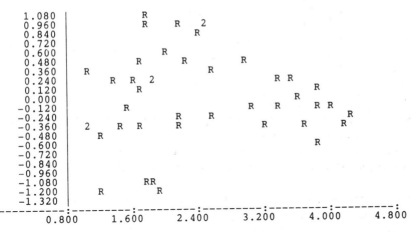

Fig. 3.9.2   Residuals v fitted values, log (SPEC) as y-variable

residual standard deviation is proportional to the prediction rather than being constant. Try typing

```
$CALCULATE LSPE = %LOG(SPEC)
$YVARIABLE LSPE              !YOU WILL GET A MESSAGE HERE
$FIT .                       !REPEAT THE PREVIOUS FIT
$CALCULATE RES = %YV -%FV    !%YV IS THE CURRENT Y VARIABLE
$PLOT RES %FV $
```

The resulting plot is shown in Fig. 3.9.2. The log transformation may have been a bit extreme as far as equalizing the residual variance goes; the larger %FVs now seem to have the smaller residuals. Accepting it for the time being, display the estimates

```
        estimate            s.e.        parameter
1        0.4925            2.316         1
2       -0.002968         0.002296       DIST
3        0.006568         0.04158        LAT
4       -0.03262          0.03698        LONG
5       -4.198e-05        0.00007484     AREA
6        0.1485           0.02694        HAB
7        0.002820         0.002380       ELEV
scale parameter taken as    0.3991
```

The LAT and LONG coefficients are still small relative to their standard errors. Take them out by typing

```
$FIT - LAT - LONG
$DISPLAY E
```

and produce

```
        estimate            s.e.        parameter
1        0.6324            0.2235        1
2       -0.002160         0.001934       DIST
3       -4.864e-05        0.00007089     AREA
4        0.1547           0.02515        HAB
5        0.002087         0.002188       ELEV
scale parameter taken as    0.3873
```

The change in deviance on removing these two variables is 0.3508 so that the F-ratio is $(0.3508/2)/0.3991 = 0.44$ with 2 and 36 d.f., still on the small side but not at all surprisingly so.

The new estimates when compared with their standard errors suggest that in fact only HAB may be worth having in the model. Try fitting HAB alone, calculate the residuals and plot them against HAB. This produces Fig. 3.9.3. A close look suggests a hint of curvature in this plot with the residuals for the larger values of HAB tending to slope downwards.

It may not be very reasonable to expect number of habitats measured directly to have a linear relationship with number of species on a log scale. Try

```
$CALCULATE LHAB = %LOG(HAB)
$FIT LHAB
$CALCULATE RES = %YV - %FV
$PLOT RES %FV $
```

and get Fig. 3.9.4. This looks rather more satisfactory.

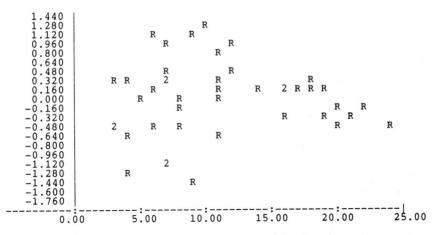

Fig. 3.9.3   Residuals v predictors, regression of log (SPEC) on HAB

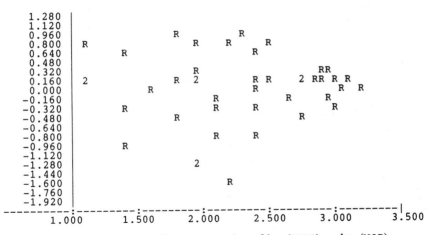

Fig. 3.9.4   Residuals v predictors, regression of log (SPEC) on log (HAB)

We might try logging the other predictors and adding them back in.

```
$CALCULATE LDIS = %LOG(DIST)
:LARE = %LOG(AREA)
:LELE = %LOG(ELEV)
$FIT + LARE + LDIS + LELE
$DISPLAY E $
```

getting

	estimate	s.e.	parameter
1	−0.6791	0.4083	1
2	0.9332	0.2597	LHAB
3	0.1603	0.06153	LARE

```
4        -0.1047        0.04755        LDIS
5         0.1404        0.1039         LELE
scale parameter taken as  0.2805
```

Elevation is still not doing much good, but the other two new *x*s look as though they were needed in the model. For a final attempt, remove LELE, display the estimates and plot the residuals. You will get

```
         estimate          s.e.       parameter
1        -0.5221        0.3955         1
2         1.048         0.2479         LHAB
3         0.1807        0.06027        LARE
4        -0.1005        0.04795        LDIS
scale parameter taken as  0.2864
```

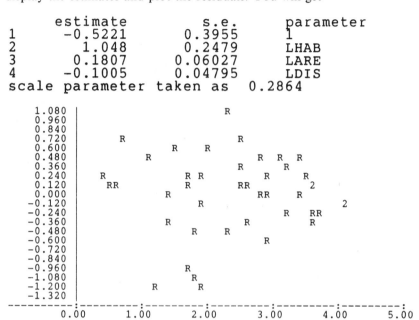

Fig. 3.9.5 Residuals v fitted values, regression of log (SPEC) on logs of HAB, AREA and DIST

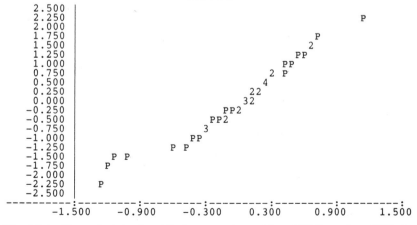

Fig. 3.9.6 Normal plot of residuals, regression of log (SPEC) on log (HAB), log (AREA) and log (ELEV)

The number of habitats seems on this evidence to be the dominant factor in determining the number of species, but having allowed for this the area and the distance from the mainland also seem to be of some importance. If you make a Normal plot of the residuals (Section 3.6) you will get the plot in Fig. 3.9.6. It looks as though there are four rather large negative residuals, so there is still something to be investigated.

Choosing a linear model is a difficult task about which a great deal has been written (there is a lot more to it than the automatic stepwise methods which are often used). Apart from the books listed in the Introduction, you should consult the article by H. V. Henderson and P. F. Velleman (1981), 'Building multiple regression models interactively'. *Biometrics* **37**, 391–412.

4

Data in tables

4.1 Models for data in tables

Suppose that in a survey we measure the level of a blood constituent and obtain the following mean values (numbers of cases in brackets):

Mean level	
Group A	Group B
118(125)	92(125)

It is apparent that the average level is lower in Group B than in Group A and we may be tempted to look for a causal factor in the environment. But suppose we look first at men and women separately. For the men we find

Group A	Group B
70(25)	80(100)

so that in men it is Group B that has the higher level. We might guess that Group B must have much lower levels in women; but in fact this is not so, the actual mean values being

Group A	Group B
130(100)	140(25)

Just as in men it is Group B that has the higher mean level. Note that this is nothing whatsoever to do with significance; if you like, you can multiply all the numbers of cases by 1000 without changing the situation.

The data, to use a nomenclature that we have met before, are classified by two *factors* each with two *levels*: sex (levels men and women) and group (levels A and B). If we make a proper tabulation of the data (Table 4.1.1), one feature stands out; the difference between the two groups is exactly the same in both men and women, and you may notice that the difference

Table 4.1.1   Blood data classification

	Group A	Group B	All	B − A
Men	70(25)	80(100)	78(125)	+ 10
Women	130(100)	140(25)	132(125)	+ 10
All	118(125)	92(125)	105(250)	− 26

between men and women is the same in both groups. When this is so we say that the two factors do not *interact*. We also say that the separate effects of the two factors are *additive*. Changing from Group A to Group B adds 10 to the means, changing from men to women adds 60, and making both changes simply adds these effects together. Notice that if there were interaction between the two factors, it would not make sense to talk about *the* difference between Group A and Group B; we would need to specify which sex we were talking about.

What is causing the trouble is the fact that (a) sex *per se* is associated with blood level, and (b) sex is very unequally distributed between the two groups, with a predominance of males in Group B and of females in Group A. Sex in this survey is a *confounding factor*. Such factors are common in medical and other surveys; we often wish to compare groups which differ as to age, sex and other factors which are known to influence the quantity we are interested in. This is a typical aspect of an *observational study*. In an experimental situation we can control some of the confounding factors and randomize the rest so that gross imbalances should not occur.

Suppose we observed the same mean values but with different numbers of cases (Table 4.1.2). The difference between the overall means for the groups

Table 4.1.2   Blood data classification

	Group A	Group B
Men	70(40)	80(60)
Women	130(60)	140(90)
All	106(100)	116(100)

is now a faithful reflection of the differences for the two sexes separately. The numbers of cases in the cells of the table are now *proportional*—the ratio of men to women is the same in each group (equivalently, the value of χ^2 for the numbers of cases is zero, so that the confounding factor, sex, and the factor of interest, group, are unassociated). Such a table is called *orthogonal*. Tables with equal numbers in the cells come under this heading. The relatively simple methods of analysis of the elementary textbooks only work for orthogonal tables, so we need methods for disentangling the more complicated non-orthogonal situations.

We can make a step forward by setting up a model for the behaviour of the data in the table. Suppose y_{ij} stands for the mean in row i and column j and that we have

$$E(y_{ij}) = \mu + \rho_i + \kappa_j$$

where μ is a constant entering into all the ys, ρ_i appears in all the ys in row i and κ_j appears in all the ys in column j. This is an additive model with no interaction and it exactly describes the sort of data we have been looking at. You can check that the four entries in the table are reproduced by taking

$$
\begin{array}{llll}
\mu & = 70 & & \\
\rho_1 & = 0 & \rho_2 & = 60 \\
\kappa_1 & = 0 & \kappa_2 & = 10
\end{array}
$$

The difference between κ_1 and κ_2 is just the between-group difference that we need to know.

With other data, how would we know if a model like this was a good model? In particular, how can we investigate whether or not we need a model involving interaction between the two factors? The answer is a very general one. We can write down a model which does allow for interaction—it might look like

$$E(y_{ij}) = \mu + \rho_i + \kappa_j + \gamma_{ij}$$

This model has more terms in it and will have a smaller residual sum of squares than that of the simpler additive model. But it will also have fewer degrees of freedom. We can now ask whether the *extra sum of squares* resulting from including the interaction term in the model is significantly large when compared to the residual mean square. This is exactly what we did in Chapter 2 when we asked whether we could drop two of the xs from the model.

In fact we can rewrite our present models so that they take the form of multiple regressions. We have y-values for the four cells in the table; let us invent some xs to go with them as shown in Table 4.1.3. These xs are such that

$$
\begin{array}{ll}
x_1 & = 1 \text{ in row 1 and 0 elsewhere} \\
x_2 & = 1 \text{ in row 2 and 0 elsewhere} \\
x_3 & = 1 \text{ in column 1 and 0 elsewhere} \\
x_4 & = 1 \text{ in column 2 and 0 elsewhere}
\end{array}
$$

Now suppose we fit the multiple regression

$$E(y) = \beta_0 + \beta_1 x_1 + \beta_2 x_2 + \beta_3 x_3 + \beta_4 x_4$$

Table 4.1.3 Blood data: multiple regression model

		Group A		Group B
Men	y	70	y	80
	x_1	1	x_1	1
	x_2	0	x_2	0
	x_3	1	x_3	0
	x_4	0	x_4	1
Women	y	130	y	140
	x_1	0	x_1	0
	x_2	1	x_2	1
	x_3	1	x_3	0
	x_4	0	x_4	1

Close inspection will show that this is exactly the same as the additive model with the following changes of notation:

$$\mu \rightarrow \beta_0$$
$$\rho_1 \rightarrow \beta_1 \qquad \rho_2 \rightarrow \beta_2$$
$$\kappa_1 \rightarrow \beta_3 \qquad \kappa_2 \rightarrow \beta_4$$

We can now use Glim to fit this model in the usual way and to tell us the residual deviance and the parameter estimates.

4.2 Fitting models to data in tables

To illustrate the method, let us look at the larger dataset in Table 4.2.1 which gives the mean birthweight of babies classified by two factors describing the smoking habits of their mothers. Looking at the bottom row you might think that smoking before pregnancy tended to decrease birthweight, but this does not show up in the individual rows of the table. It is clear that the factors are far from orthogonal—most heavy smokers before pregnancy went on smoking during pregnancy. Can we clarify the situation?

The logical first step is to look for interaction between the two factors, and we do this by fitting two models as described above. You should put the data in a file in the layout

```
7.66  252    7.68   50    7.59   19
7.22  187    7.42   43    7.45   11
7.13  106    7.34   64    7.08   26
7.14   76    7.19  124    7.17  138
```

and type

```
$UNITS 12    $DATA BWT N    $DINPUT 8 $
```

Table 4.2.1 Birthweights (lb) of babies by smoking of mothers before and during pregnancy

Cigarettes smoked during pregnancy	Cigarettes smoked before last pregnancy			
	< 32000	32000–63999	64000 +	Combined
None	7.66(252)	7.68(50)	7.59(19)	7.66(321)
< 2000	7.22(187)	7.42(43)	7.45(11)	7.26(241)
2000–3999	7.13(106)	7.34(64)	7.08(26)	7.19(196)
4000 +	7.14(76)	7.19(124)	7.17(138)	7.13(338)
Combined	7.37(621)	7.30(281)	7.21(194)	7.32(1096)
	Within cell s.d. 0.67 (1084 d.f.)			

to read them in. Instead of typing in the factor levels it is quicker to type

```
$CALCULATE ROWS = %GL(4,3)    : COLS = %GL(3,1) $
```

There are now some preliminary steps to take. First, the entries in the table are means, not single observations. To take account of this we need to use *weighted regression* with the numbers in the cells as weights. This requires the directive

```
$WEIGHT N $
```

We are told that the within-cell standard deviation is 0.67, so we should specify that the scale parameter is to be taken to be the square of this—

```
$CALCULATE %A = 0.67 ** 2    $SCALE %A $
```

Most importantly, we need to specify that ROWS and COLS are not measurements but factors (recall Section 2.4). We want Glim to create the 'dummy' xs that enable us to use multiple regression methods, and to do this you should type

```
$FACTOR ROWS 4 COLS 3 $
```

where the numbers are the numbers of levels to be allowed for. Finally you must remember to specify the y-variable:

```
$YVARIABLE BWT $
```

Now we can go ahead. First let us fit the additive model without interaction. Type

```
$FIT ROWS + COLS    $DISPLAY E $
```

displaying the estimates for future reference. This produces

```
scaled deviance = 4.6602
         d.f. = 6
```

```
       estimate          s.e.        parameter
1         7.640        0.03874       1
2        -0.3948       0.05712       ROWS(2)
3        -0.4884       0.06181       ROWS(3)
4        -0.5240       0.05936       ROWS(4)
5         0.1097       0.05166       COLS(2)
6         0.03577      0.06331       COLS(3)
scale parameter taken as  0.4489
```

If you remember Section 2.4, you will not be surprised by the absence of estimates for ROWS(1) and COLS(1), both of which are set to equal 0 ± 0.

Now let us fit a model with interaction between the two factors and see what happens to the deviance. The Glim notation for the interaction between two factors A and B is A.B, so we need to type

```
$FIT +ROWS.COLS
```

The result may be a bit of a surprise. It will depend upon just what computer you are using, but will look something like

```
scaled deviance = 9.085e-28 (change =  -4.660)
         d.f. =      0       (change =  -6     )
```

(Some implementations may accompany this by some frightening messages about floating-point underflow. Ignore them.)

If you think back to the details of the model with interaction, it says that the column pattern may differ arbitrarily from row to row. This means that the model imposes no restriction whatever upon the pattern of the fitted cell means. Consequently they are capable of reproducing the observed cell means exactly, and this is just what has happened. The tiny residual deviance is the machine's best shot at an exact zero, and you may have noted that it has zero d.f. All this need cause you no worry; we can still write down the analysis of scaled deviance so far (Table 4.2.2).

Table 4.2.2 Analysis of scaled deviance

	d.f.	Scaled deviance	Mean scaled deviance
Model without interaction	6	4.6602	
Model with interaction	0	0	
Extra due to interaction	6	4.6602	0.7767
Within cells	1084		1.0000

The F-ratio is less than 1.0 and this says that the departures of the cell means from the values predicted by the additive model are on the whole slightly smaller than the within-cell variability would lead you to expect. As far as this is concerned, the additive model is an excellent fit to the data.

Having got so far, we can go on to ask some further questions. Do we need both factors to explain the data or will just a single factor do? We answer this just as before by fitting the competing models and looking at the changes in deviance. If you type

```
$FIT ROWS    $
```

you will get a scaled deviance of 9.2537 on 8 d.f. This gives the analysis shown in Table 4.2.3. The F-ratio of 2.2968 gives $p \cong 0.10$. Fitting columns alone leads to the analysis in Table 4.2.4. With this big F-ratio the significance probability is vanishingly small.

Table 4.2.3   Analysis of scaled deviance

	d.f.	Scaled deviance	Mean scaled deviance
Model with rows only	8	9.2537	
Model with rows and columns	6	4.6602	
Extra due to columns	2	4.5935	2.2968
Within cells	1084		1.0000

Table 4.2.4   Analysis of scaled deviance

	d.f.	Scaled deviance	Mean scaled deviance
Model with columns only	9	107.7864	
Model with rows and columns	6	4.6602	
Extra for rows	3	103.1262	34.3754
Within cells	1084		1.0000

The upshot of all this is that there are very significant differences between the levels of the row factor (smoking during pregnancy) after allowing for any possible differences between columns. The differences between the levels of the column factor (smoking before pregnancy) are of doubtful significance once smoking during pregnancy has been taken into account.

4.3 Standard errors

Of course, these statements are quite inadequate as a summary of the findings of the analysis; it is essential to look at the estimates of the parameters. Harking back to the display following the fit of the additive model, these were

1	7.640	ROWS(1)	0	COLS(1)	0
		(2)	−0.3948	(2)	0.1097
		(3)	−0.4884	(3)	0.0358
		(4)	−0.5240		

Note that, if anything, the first column has the smallest expected birth-weight, not the largest.

What standard errors should we use for comparing these estimates? This gets a little complicated. It is quite easy to make comparisons with level 1 of a factor since the corresponding 'estimate' is 0 ± 0. This means that the display entry $+0.1097\pm0.0517$ for COL(2) is actually the difference between column 2 and column 1 along with the standard error for this difference. The value of t is $0.1097/0.0517 = 2.12$, just outside the 5% point.

To get the other standard errors we need some further information. Glim will print out the standard errors of all the differences if you type

$DISPLAY S $

(You will need to re-fit the rows + columns model first.) The output will be

```
S.E.s of differences of parameter estimates
1       0.00
2       0.0869      0.00
3       0.0888      0.0654      0.00
4       0.0855      0.0632      0.0629      0.00
5       0.0718      0.0776      0.0871      0.0907      0.00
6       0.0799      0.0850      0.0941      0.104       0.0640
          1           2           3           4           5
6       0.00
          6
     scale parameter taken as   0.449
```

and we see that, for example, the standard error of the difference between columns 2 and 3 (i.e. between the estimates of parameters 5 and 6 in the E display) is ±0.0640.

This is all right for simple differences, but we might be interested in some more complicated functions of the estimates. Staying with the column factor, what if we compare the zero 'estimate' in column 1 with the mean of the estimates for columns 2 and 3? This difference is $(0.1097 + 0.0358)/2 = 0.0728$ and to get its standard error we need the variances and covariance of the constituent estimates. You can get these by typing

$DISPLAY V $

which will produce

```
(Co)variances of parameter estimates
1      0.00150
2     -0.00139      0.00326
3     -0.00128      0.00140      0.00382
4     -0.00114      0.00140      0.00169      0.00352
5     -0.000492    -4.30e-05    -0.000550    -0.00101      0.00267
6     -0.000438     0.0000251   -0.000515    -0.00167      0.00129
         1            2            3            4            5
6      0.00401
         6
     scale parameter taken as   0.449
```

(a few more significant figures would be desirable). The standard error we want is now found as

$$\sqrt{(0.00267 + 2 \times 0.00129 + 0.00401)/4} = 0.0481$$

4.4 An example with three factors

Table 4.4.1 gives the data from an experiment with three factors. The quantity measured is the log of the ratio of the adrenal weight to total body weight in rats, and the animals are classified by female parent (FP, 3 levels), male parent (MP, 3 levels) and sex (SEX, 2 levels). For each of the $2 \times 3 \times 2 = 18$ factor-level combinations there are 2 animals but three of the readings are missing.

Table 4.4.1   Relative adrenal weights of rats (log scale)

					Female parent				
			1		2			3	
Male parent	1	F	0.97	1.23	F	1.25	1.20	F	1.16	1.
		M	0.84	0.92	M	–	0.86	M	0.94	0."
	2	F	1.15	1.29	F	1.27	1.23	F	1.33	1.
		M	0.70	0.87	M	0.86	0.82	M	1.00	0."
	3	F	1.35	1.37	F	–	–	F	1.21	1.
		M	0.93	0.99	M	1.02	0.98	M	0.91	0.

As before we need to consider models in which the effect of SEX (the male–female differential) varies from one level of FP or MP to another. These models contain two-factor interactions like FP.SEX and MP.SEX. But we can also envisage a situation where SEX interacts with FP but to a different extent depending on the level of the third factor, MP. If this occurs we say that there is a three-factor interaction and write it as FP.MP.SEX.

The easiest way to enter the data is to put them into a file using the following layout:

```
0.97   1.23   0.84   0.92
1.25   1.20   0.00   0.86
1.16   1.28   0.94   0.98
1.15   1.29   0.70   0.87
1.27   1.23   0.86   0.82
1.33   1.34   1.00   0.98
1.35   1.37   0.93   0.99
0.00   0.00   1.02   0.98
1.21   1.21   0.91   0.76
```

putting a zero in place of each of the missing readings. You can then type

```
$UNITS 36 $  $DATA ADRE   $DIN 8 $
$FACTOR FP 3 MP 3 SEX 2
$CALCULATE FP = %GL(3,4)   : MP = %GL(3,12)   :SEX = %GL(2,2)
```

To eliminate the missing values from the analysis we can use a weighted regression with weights equal to 0 for the missing values and to 1 for the rest. You can construct the weights by typing

```
$CALCULATE W = 1   : W(7) = W(29) = W(30) = 0
$WEIGHT W
$YVARIABLE ADRE $
```

The first step in the analysis is to get the within-cell mean square to use as a scale parameter. With data classified by three factors we can fit the cell means exactly by a model including all possible interactions—we could type

```
$FIT FP + MP + SEX + FP.MPO + FP.SEX + MP.SEX + FP.MP.SEX $
```

However, Glim allows you to abbreviate this to

```
$FIT FP * MP * SEX $
```

This will be found to give you a deviance of 0.086400 on 16 d.f., so next type

```
$CALCULATE %A = %DV/%DF     $SCALE %A $
```

(you will remember that %DV and %DF are system scalars equal to the deviance and its degrees of freedom).

Now we can start to investigate alternative models. It would be nice if we could do without the three-factor interaction. Try

```
$FIT -FP.MP.SEX $
```

This gives a scaled deviance of 17.3657 on 19 d.f. and an analysis of deviance as shown in Table 4.4.2. The F-ratio is less than 1.0 and there seems no good

Table 4.4.2 Analysis of scaled deviance

	d.f.	Scaled deviance	Mean scaled deviance
Model with all interactions	16	16.0000	1.0000
Model without 3-factor interaction	19	17.3657	
Extra for 3-factor interaction	3	1.3657	0.4552

reason to retain the complex interaction in the model. (Those familiar with the analysis of variance may have expected $(3-1) \times (3-1) \times (2-1) = 4$ d.f.

for the interaction between factors with 3, 3 and 2 levels. The missing d.f. is due to the fact that one whole cell is missing from the complete three-way layout.)

Continuing our exploration, try leaving out one of the two-factor interactions. Omitting FP.MP leads to the analysis shown in Table 4.4.3. The

Table 4.4.3   Analysis of scaled deviance

	d.f.	Scaled deviance	Mean scaled deviance
Main effects, FP.SEX, MP.SEX	23	39.8239	
Main effects,			
all 2-factor interactions	19	17.3657	
Extra for FP.MP	4	22.4582	5.6145
Within cells	16		1.0000

F-value is significant at $p = 0.01$ so that we cannot safely do without this interaction; the differences between female parents are not the same for all three male parents. We had better put it back before going on. On the other hand, it turns out that FP.SEX and MP.SEX do not produce large extra deviances. If we type

```
$FIT +FP.MP - FP.SEX - MP.SEX   $
```

we get Table 4.4.4. The *F*-ratio gives $p \cong 0.25$ and we can reasonably omit these interactions from the model.

Table 4.4.4   Analysis of scaled deviance

	d.f.	Scaled deviance	Mean scaled deviance
Main effects, FP.MP	23	23.2448	
Main effects,			
all 2-factor interactions	19	17.3657	
Extra for FP.SEX, MP.SEX	4	5.8790	1.4698
Within cells			1.0000

What about the main effects? Technically we could look at the SEX main effect by omitting SEX from the model, but the female readings are clearly a good deal bigger than the male ones and a formal test of significance is rather

waste of time. If we try leaving out FP or MP we may get rather a surprise.
Omitting them both for the sake of illustration, type

$FIT - FP - MP $

This gives a deviance of 23.2448 on 23 d.f., exactly the same as before. The
point is rather similar to one we made earlier; the model with two main
effects and their interaction prescribes no more than an arbitrary pattern of
values for the corresponding two-way table, and this is equally well
described by a model like

$$E(y_{ijk}) = \mu + \alpha_k + \gamma_{ij}$$

which consists of terms for SEX and FP.MP only. Try typing

$DISPLAY E
$FIT + FP + MP $DISPLAY E $

nd compare the two sets of parameter estimates. They are

```
      estimate            s.e.       parameter
1        1.165          0.03907      1
2       -0.3507         0.02654      SEX(2)
3        0.01250        0.05196      FP(1).MP(2
4        0.1700         0.05196      FP(1).MP(3
5        0.05489        0.05630      FP(2).MP(1
6        0.05500        0.05196      FP(2).MP(2
7        0.1853         0.06501      FP(2).MP(3,
8        0.1000         0.05196      FP(3).MP(1)
9        0.1725         0.05196      FP(3).MP(2)
10       0.03250        0.05196      FP(3).MP(3)
scale parameter taken as   0.005400
```

nd

```
      estimate            s.e.       parameter
1        1.165          0.03907      1
2        0.05489        0.05630      FP(2)
3        0.1000         0.05196      FP(3)
4        0.01250        0.05196      MP(2)
5        0.1700         0.05196      MP(3)
6       -0.3507         0.02654      SEX(2)
7       -0.01239        0.07661      FP(2).MP(2)
8       -0.03957        0.08668      FP(2).MP(3)
9        0.06000        0.07348      FP(3).MP(2)
10      -0.2375         0.07348      FP(3).MP(3)
scale parameter taken as   0.005400
```

You should check (by hand calculation or by typing $DISPLAY R $) that these
two sets of estimates give exactly the same fitted values—they express the
same model in two different forms. The one without the main effects is
slightly easier to interpret since the eight FP.MP items correspond to the
differences between the mean of the (1,1) FP.MP cell and the eight other cell

means. As usual you can type

```
$DISPLAY S $
```

to get the standard errors of the differences.

Our final model, then, contains the SEX main effect and the two-way table for FP by MP. The method of presenting the results should reflect this but is to some extent a matter of taste. For the SEX effect you can display the grand mean plus and minus half the SEX(2) estimate:

Females	Males	Difference
1.340	0.990	-0.350 ± 0.027

For the two-way table it is easiest to type $DISPLAY R $ to get the fitted 'values (if you have not done so already) and then to take the mean of the female and male figures in each cell. This gives Table 4.4.5 with standard errors of differences ranging from 0.164 to 0.237. Note that to do this we must include the fitted value in the empty cell of the table.

Table 4.4.5   Means for male and female parents

		MP 1	2	3
	1	0.990	1.045	1.090
FP	2	1.003	1.045	1.163
	3	1.160	1.176	1.023

4.5 Aliasing

While we are dealing with this example, try typing

```
$FIT FP.MP.SEX      $DISPLAY E      $
```

This should reproduce the cell means, but what happens to the missing one? The output is

```
       estimate        s.e.       parameter
1        1.100       0.05196      1
2       -0.2200      0.07348      FP(1).MP(1).SEX(2
3        0.1200      0.07348      FP(1).MP(2).SEX(1
4       -0.3150      0.07348      FP(1).MP(2).SEX(2
5        0.2600      0.07348      FP(1).MP(3).SEX(1
6       -0.1400      0.07348      FP(1).MP(3).SEX(2
7        0.1250      0.07348      FP(2).MP(1).SEX(1
8       -0.2400      0.09000      FP(2).MP(1).SEX(2
```

```
 9       0.1500      0.07348     FP(2).MP(2).SEX(1)
10      -0.2600      0.07348     FP(2).MP(2).SEX(2)
11       0.000       aliased     FP(2).MP(3).SEX(1)
12      -0.1000      0.07348     FP(2).MP(3).SEX(2)
13       0.1200      0.07348     FP(3).MP(1).SEX(1)
14      -0.1400      0.07348     FP(3).MP(1).SEX(2)
15       0.2350      0.07348     FP(3).MP(2).SEX(1)
16      -0.1100      0.07348     FP(3).MP(2).SEX(2)
17       0.1100      0.07348     FP(3).MP(3).SEX(1)
18      -0.2650      0.07348     FP(3).MP(3).SEX(2)
scale parameter taken as   0.005400
```

The missing cell has produced a zero estimate labelled 'aliased'. You should note that this corresponds exactly to the missing level 1 entries that you are now used to—these could have been included and labelled in just the same way. The 'aliased' message always indicates that we are trying to estimate more parameters than the structure of the data will permit. The interpretation can be less straightforward. If you fit the same model yet again, this time by typing

```
$FIT FP*MP*SEX    $DIS E    $
```

you will get

```
        estimate          s.e.      parameter
 1        1.100         0.05196     1
 2        0.1250        0.07348     FP(2)
 3        0.1200        0.07348     FP(3)
 4        0.1200        0.07348     MP(2)
 5        0.2600        0.07348     MP(3)
 6       -0.2200        0.07348     SEX(2)
 7       -0.09500       0.1039      FP(2).MP(2)
 8        0.06000       0.1162      FP(2).MP(3)
 9       -0.005000      0.1039      FP(3).MP(2)
10       -0.2700        0.1039      FP(3).MP(3)
11       -0.1450        0.1162      FP(2).SEX(2)
12       -0.04000       0.1039      FP(3).SEX(2)
13       -0.2150        0.1039      MP(2).SEX(2)
14       -0.1800        0.1039      MP(3).SEX(2)
15        0.1700        0.1559      FP(2).MP(2).SEX(2)
16        0.000         aliased     FP(2).MP(3).SEX(2)
17        0.1300        0.1470      FP(3).MP(2).SEX(2)
18        0.06500       0.1470      FP(3).MP(3).SEX(2)
scale parameter taken as   0.005400
```

It is somewhat less obvious just why parameter 16 is aliased, though the fact does indicate why there are only 3 degrees of freedom for the 3-factor interaction.

It is possible (though usually undesirable) to include the aliased parameters in the computation by typing

```
$ALIAS
```

This will lead to the last level of each factor being aliased instead of the first.

5

Tabulation (Glim77 only)

@ @ @ @

5.1 Tabulating data

Glim77 has a $TABULATE directive which has a rather different format from
the ones we are used to by now. It operates through a set of *phrases*, each
phrase being introduced by a *keyword*.

For the whole of this chapter we are going to use the Clinic data set from
Table 5.1.1 to illustrate the commands. You need to put this into a file and
read it in by commands such as

```
$UNITS 41
$DATA SEX AGE SC WEIGHT DIET SMOKER ALCOHOL
$DINPUT 8 $
```

Note that the − 1s in this dataset denote unknown values.

In the simplest case the output from the $TABULATE directive is a no-way
table, that is to say a single number which is the value of some statistic
relating to a particular variable. The necessary keyword is THE (!) and this
has to be followed by the name of the variable and that of the required
statistic. Thus you could type

```
$TABULATE THE AGE MEAN $
```

which produces the number 53.59, the mean of the age values. The possible
statistics are

Mean
Total
Variance
Deviation (i.e. standard deviation)
Smallest
Largest
Fifty (i.e. median)
Percentile
Interpolate

The peculiar names are due to the fact that only the initial letter of the
statistic name matters. The names are fairly self-explanatory except perhaps
for the last three. The Percentile and Interpolate names must each be

60

Table 5.1.1 Clinic dataset

Sex	Age	S.C.	Wt	Diet	Smok	Alcohol
2	54	4	56	1	1	100
1	44	2	89	1	2	120
1	59	3	55	1	1	130
1	55	3	66	1	2	100
2	68	4	58	2	1	80
2	43	3	−1	2	2	200
2	72	3	−1	2	1	220
2	51	5	53	1	1	100
2	57	3	−1	2	1	120
2	43	3	−1	1	1	120
1	37	2	67	1	3	180
1	64	1	80	1	1	120
1	45	3	81	1	1	−1
2	62	3	−1	1	3	−1
2	42	2	44	1	1	−1
1	51	1	81	1	2	150
2	57	3	68	1	1	−1
1	55	3	62	1	1	−1
2	34	4	70	1	1	200
2	73	4	51	2	3	−1
2	72	3	40	1	1	−1
1	70	3	55	1	3	100
2	44	3	96	1	1	−1
2	58	2	48	1	1	130
2	40	3	−1	1	1	−1
2	51	2	64	1	1	−1
2	52	3	50	1	1	−1
2	54	3	−1	1	2	126
1	40	4	66	3	1	170
1	60	2	10	2	1	220
2	46	2	71	2	1	180
1	63	2	77	2	1	120
2	40	3	53	2	1	300
2	54	3	−1	1	1	−1
2	77	1	55	3	1	120
2	52	1	44	3	3	−1
1	39	4	60	1	1	200
1	53	2	−1	1	3	120
1	55	2	90	1	1	85
1	48	2	85	1	1	−1

Table 5.1.1 *Continued*

Sex	Age	S.C.	Wt	Diet	Smok	Alcohol
1	55	1	65	1	3	-1

Column 1—Sex (1 = male, 2 = female)
          2—Age (years)
          3—Social class
          4—Weight (kg) (-1 = unknown)
          5—Diet (1 = adequate, 2 = inadequate, 3 = unknown)
          6—Smoker (1 = yes, 2 = no, 3 = unknown)
          7—Alcohol consumption (g/day) (-1 = unknown)

followed by a number between 0 and 100 to specify which percentile or interpolate is needed. A percentile is always equal to a variable value or to the average of two adjacent values; an interpolate is linearly interpolated between adjacent values. Suppose for instance that the variable X takes the values 1, 2, ..., 10. Then the 25 percentile and interpolate are both equal to 3.0 and the 30 percentile and interpolate are both equal to 3.5; but the 27 percentile is equal to 3.0 whereas the 27 interpolate is equal to 3.2. The name Fifty is equivalent to Percentile 50.

To produce a table classified by one or more factors, a FOR keyword is used followed by the factor names separated by semicolons. You could type

$TABULATE THE AGE MEAN FOR SEX;SC $

and get the table below:

```
          1.000   2.000   3.000   4.000   5.000
1.000    56.67   51.43   56.80   39.50    0.00
2.000    64.50   49.25   53.08   57.25   51.00
```

If there is no THE phrase, a table of counts is produced, so that

$TABULATE FOR SEX;SC $

produces

```
          1.000   2.000    3.000   4.000   5.000
1.000    3.000   7.000    5.000   2.000   0.000
2.000    2.000   4.000   13.000   4.000   1.000
```

Note that the names in the FOR phrase do not have to be those of factors in the technical Glim sense; the distinct values of ordinary variables are treated as factor levels for this purpose.

With the phrases used so far the table will simply be displayed, but its values can also be stored in a variable for future reference. This requires an

INTO phrase giving the name of the vector in which the table values are to be stored. The order in which the cell values are stored is governed by the order of the factor names in the FOR phrase, with the levels of the last-quoted factor moving fastest. For example, if the factors of a three-way table have 2, 3 and 2 levels, the output will be in the sequence

111 112 121 122 131 132 211 212 221 222 231 232

It is also possible to produce new variables containing the factor levels which index the table. This requires a BY phrase. The levels of a Glim factor are simply the integers 1, 2, ... The 'levels' of a variable are its distinct values in ascending order. The BY keyword must be followed by variable names separated by semicolons, specifying as many variables as there were in the FOR phrase. Alternatively, names of scalars can be given; in this case, the numbers of levels of the corresponding factors will be stored. If you type

```
$TABULATE THE WEIGHT MEAN FOR SMOK BY SLEV
$LOOK SLEV $
```

you will get

```
SLEV   1.000    2.000    3.000
  [ ]   52.75   56.00    40.00
        SLEV
1       1.000
2       2.000
3       3.000
```

Note that when an INTO phrase causes a table to be stored, it will not be automatically printed. We shall see how to print a stored table in the next section.

5.2 Printing tables

The contents of any vector can be displayed in tabular form by the $TPRINT command. The layout of the command is fairly simple; you have to specify the name of the vector which contains the table values and follow this by the numbers of levels of the factors separated by semicolons. Thus the commands

```
$UNITS 24
$CALCULATE X = %CU(1)      !GENERATE INTEGERS FROM 1 TO 24
$TPRINT X 3;4;2 $
```

gives the output

```
             1         2
 1  1     1.000     2.000
    2     3.000     4.000
    3     5.000     6.000
    4     7.000     8.000
```

```
2 1    9.000   10.000
  2   11.000   12.000
  3   13.000   14.000
  4   15.000   16.000

3 1   17.000   18.000
  2   19.000   20.000
  3   21.000   22.000
  4   23.000   24.000
```

The $TPRINT command tries to output the table as compactly as possible. Rather than quote its rather complicated rules (which you can find in the Glim77 *Users' Guide*) I suggest you try displaying a variety of tables with different numbers of factors and levels and see what happens.

When $TPRINT is used to display the values in a vector X, these values are assumed to be in *standard order* with respect to the factor levels. This means that the levels of each factor form a nested set from left (changing slowest) to right (changing fastest), with the levels of a factor increasing within each level of the surrounding one. This is the order in which the table is stored by the $TABULATE INTO command as we saw in the previous section. Hence if we wish to store and display a table of means, for example, we might type something like

```
$TABULATE THE AGE MEAN FOR SEX;SC INTO TAB
$TPRINT TAB 2;5 $
```

producing the same table as before.

You can also specify the table layout by including explicit factor levels. Suppose we followed the original commands in this section by the following:

```
$CALCULATE F1 = %GL(3,8)   : F2 = %GL(4,2)   : F3 = %GL(2
$FACTOR F1 3   F2 4   F3 2
$TPRINT X F1;F2;F3 $
```

the result will be exactly as before except that the factor names will be included in the table. However, you could also rearrange the layout of the table by typing

```
$TPRINT X F3;F2;F1
```

This would produce

```
     F1     1         2         3
F3   F2
 1    1    1.000     9.000    17.000
      2    3.000    11.000    19.000
      3    5.000    13.000    21.000
      4    7.000    15.000    23.000

 2    1    2.000    10.000    18.000
      2    4.000    12.000    20.000
      3    6.000    14.000    22.000
      4    8.000    16.000    24.000
```

Note that with this form of the $TPRINT command it is only necessary for the table values to be in standard order with respect to some arrangement of the factors.

Just as with the $TABULATE command the classifying factors of the printed table need not be factors in the technical Glim sense. If variables are used, the table will be labelled with their names and their actual distinct values.

If you specify more than one variable name (separated by semicolons) in the first part of the $TPRINT command, values belonging to each of the variables will appear in each cell of the table. Thus

```
$TABULATE THE AGE MEAN FOR SEX;SC INTO TAB
: THE AGE DEVIATION FOR SEX;SC INTO TSD
$TPRINT TAB;TSD 2;5$
```

produces

		1	2	3	4	5
1	TAB	56.67	51.43	56.80	39.50	0.00
	TSD	6.6583	9.1078	9.0111	0.7071	0.0000
2	TAB	64.50	49.25	53.08	57.25	51.00
	TSD	17.6777	6.8981	11.0186	17.4619	0.0000

5.3 Weights

You can specify a vector of *input weights* in the $TABULATE command by means of a WITH phrase giving the vector name. A value associated with an input weight equal to n is then treated roughly as if it occurred n times. With the THE phrase calling for a total, mean, variance or standard deviation, an ordinary weighted version of the statistic is produced. If there is no THE phrase, the weights are still treated as frequencies and can be tabulated as such. With the Clinic dataset, to get the mean of the alcohol figures omitting the unknown values we might type

```
$CALCULATE W = (ALCO > 0)
$TABULATE THE ALCO MEAN WITH W $
```

obtaining the result 146.6. Counting the non-unknown values can be done by typing

```
$TABULATE WITH W $
```

One result of a tabulation is a set of *output weights*, one for each cell of the table. For variances and standard deviations these are degrees of freedom; otherwise they are counts or sums of input weights. They can be stored in a vector by means of a USING phrase giving the vector name. To print the output weights instead of storing them type the phrase USING []. Typing

```
$TABULATE THE AGE MEAN FOR SEX;SC USING [] $
```

produces

		1.000	2.000	3.000	4.000	5.000
1.000	[ ]	56.67	51.43	56.80	39.50	0.00
	[ ]	3.000	7.000	5.000	2.000	0.000
2.000	[ ]	64.50	49.25	53.08	57.25	51.00
	[ ]	2.000	4.000	13.000	4.000	1.000

5.4 Grouping and mapping

When making a table with a variable such as age it is often necessary to produce a grouped version to use as a factor. Glim77 has two directives for this purpose.

The $GROUP command is most easily described by an example. Suppose we type

```
$ASSIGN BDRY = 30, 40, 50, 60
$GROUP GAGE = AGE   INTERVALS BDRY * $
```

If an AGE value is in the interval $30 \leqslant AGE < 40$, the corresponding value in GAGE will be 1; if AGE is in $40 \leqslant AGE < 50$, GAGE will be 2; and so on. The final asterisk in the INTERVALS phrase shows that AGE values $\geqslant 60$ will give GAGE the value 4. An initial asterisk can be used to call for an 'infinite' group at the lower end of the BDRY values in the same way. Every item in AGE must be catered for; with the command shown, an AGE of 25 would be an error. The items in BDRY must be in strictly increasing order. The result GAGE is automatically declared to be a Glim factor with (in this case) four levels.

An exactly similar command is $MAP, which produces a variable instead of a factor. This usually requires a VALUES phrase giving the quantities which are to correspond to each of the grouping intervals. Thus we might go on to type

```
$ASSIGN MPTS = 35, 45, 55, 70
$MAP MAGE = AGE   INTERVALS BDRY *   VALUES MPTS $
```

With the Clinic dataset the first few values of GAGE would be

3 2 3 3 4 2 4 3 ...

while those of MAGE would be

55 45 55 55 70 45 70 55 ...

The $GROUP command can also contain a VALUES phrase; this should specify a suitable set of small integers. In the keywords INTERVALS and VALUES, only the initial letters matter.

@@@@

6

Models with factors and variables as predictors

6.1 Parallel lines

We have seen how we can use linear models with continuous x-variables and others with dummy xs which correspond to the levels of a factor or combinations of levels in an interaction term. As far as the theory and arithmetic are concerned, these are all just x-variables and there is no reason why you should not consider models containing a mixture of the two types. In this chapter we shall study the uses of some models of this type. In such models the continuous xs are often called *covariates*.

As an example, Table 6.1.1 shows measurements of blood pressure (BP) in the arm and leg of two groups of surgical patients, one anaesthetized using curare and one without this agent. The object is to see whether leg BP can be used in place of the usual arm BP during operations on the upper body. In order to predict arm BP from leg BP we might calculate a linear regression and the question is whether we can use the same equation for prediction in the two groups.

Input the data as follows:

```
$UNITS 23    $DATA ARM LEG DRUG    $READ
  95   95 1
 105  130 1
    . . . .
    . . . .
  50   80 2
  75   70 2
$YVARIABLE ARM
$FACTOR DRUG 2 $
```

The third column is a drug code with 1 for no curare and 2 for curare.

The first question that arises is whether we can use the same slope in both the groups. This is a question about an interaction between the factor DRUG and the continuous measurement LEG; is the effect of a given change in LEG (i.e. the slope of the line) the same at both levels of DRUG? Glim allows you to specify an interaction term containing one or more factors plus just one continuous measurement variable, so you can type

```
$FIT LEG + DRUG + LEG.DRUG $
```

Table 6.1.1   Blood pressures (mm Hg) in surgical patients

No curare		Curare	
Arm	Leg	Arm	Leg
95	95	125	145
105	130	100	120
120	140	105	130
95	110	110	130
95	115	70	95
70	100	65	90
85	110	70	100
95	115	80	105
85	110	95	125
110	110	55	70
		75	100
		50	80
		75	70

or just

```
$FIT LEG*DRUG $
```

This will produce a deviance of 1772.6 on 19 d.f. Set this as the scale parameter in the usual way:

```
$CALCULATE %A = %DV/%DF    $SCALE %A $
```

If we now omit the interaction term by typing

```
$FIT - LEG.DRUG $
```

the scaled deviance goes to 19.145 on 20 d.f. The extra deviance is 0.145 on 1 d.f., far below the expected value of 1.0, and the model without interaction fits the data satisfactorily. This model fits the same slope to each group; in other words, the model is one for parallel lines.

The next question is whether the intercepts of the two parallel lines can be taken to be the same, and we can re-phrase this question by asking whether we need the DRUG term in the model. Typing

```
$FIT - DRUG $
```

gives a scaled deviance of 20.756 on 21 d.f. The extra scaled deviance is 1.611 on 1 d.f., a bit greater than 1.0 but not to any significant extent. It appears that the data can be fitted by two parallel lines which coincide, i.e. by a single line.

The successive models we have considered can be illustrated by the diagrams in Fig. 6.1.1. We can also summarize the deviances as in Table

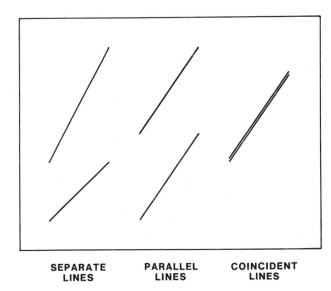

SEPARATE   **PARALLEL**   **COINCIDENT**
  LINES   **LINES**   **LINES**

Fig. 6.1.1   Alternative models for two groups

Table 6.1.2   Analysis of scaled deviance

	d.f.	Scaled deviance	Mean scaled deviance
Separate lines	19	19.000	1.000
Parallel lines	20	19.145	
Extra for parallelism	1	0.145	0.145
Coincident lines	21	20.756	
Extra for coincidence	1	1.611	1.611

6.1.2. If you display the estimates you will find that the intercept is well below its standard error, so that a line through the origin might do. Type

```
$FIT -1                    !OR -%GM WITH GLIM3
$DISPLAY E $
```

and get

```
scaled deviance = 21.115 (change = +0.3585)
          d.f. = 22       (change = +1      )

        estimate        s.e.      parameter
   1      0.8156      0.01827       LEG
   scale parameter taken as  93.29
```

This suggests that an equation

arm BP = leg BP × 0.8156

could be used for both the groups. Obviously a proper investigation of the question would need considerably larger samples.

6.2 Offsets

We have arrived at a model for our two measurements of the form $y = kx$, implying that one is proportional to the other. An obvious alternative that you might want to consider is a model involving a constant difference between the two, giving an equation $y = x + d$. There is a single unknown in this equation, the quantity d, and it seems clear that we should be able to estimate it by linear modelling methods, but we need to force the x term into the model with its coefficient fixed at 1.0. We can do this by declaring this term to be an *offset*—type.

```
$OFFSET LEG $
```

Now all that we need to fit is the constant term or intercept; typing

```
$FIT    $DISPLAY E $
```

produces

```
scaled deviance = 22.230
          d.f. = 22

          estimate        s.e.      parameter
    1      −20.22         2.014      1
    scale parameter taken as  93.29
```

suggesting a model in which arm BP is some 20 mm lower than leg BP. The scaled deviance is marginally larger than that of the proportional model, but it would take a good deal more data and a close study of the residuals to distinguish convincingly between the two.

6.3 Covariance adjustment

The data in Table 6.3.1 give a slightly different application of the same technique. Here we have heights and shoulder widths of a sample of army cadets and one of university students. The cadets have slightly broader shoulders on average, but they also have smaller heights, and shoulder width and height are correlated. What is the estimated difference in shoulder width between a cadet and a student with the same height? We can answer this question by fitting two regression lines to the data (Fig. 6.3.1) and finding the vertical distance between them. Note that the form of the question takes it for granted that the vertical distance is the same for any value of height, i.e.

Table 6.3.1 Heights and shoulder widths (cm) of army cadets and university students

	Cadets		Students	
Height	Shoulder Width	Height	Shoulder Width	
163	38.5	169	37.1	
163	39.9	169	38.3	
171	38.5	170	40.3	
171	40.5	172	36.4	
171	42.0	178	42.0	
173	40.4	179	39.9	
175	39.8	179	41.1	
176	40.5	180	38.8	
176	41.6	183	39.7	
177	41.3	187	42.9	
178	43.5			
179	43.9			
180	42.1			
180	42.3			

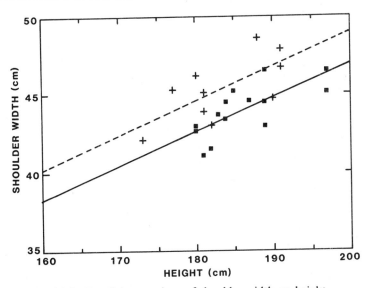

Fig. 6.3.1 Parallel regressions of shoulder width on height

that the two regression lines are parallel.

Our tactics are the same as before. Type in the data as variables HT and SHOULDER, plus the groups coded 1 and 2 in GP; define GP to be a factor with 2 levels and SHOULDER to be the *y*-variable:

```
$UNITS 24    $DATA HT SHOULDER GP    $READ
163 38.5 1
163 39.9 1
    ....
    ....
187 42.9 2

$FACTOR GP 2    $YVARIABLE SHOULDER $
```

Before the actual fitting it is worth noting that, with two groups and one covariate, all the extra deviances we are going to look at have 1 d.f. This means that the implied F-tests are equivalent to t-tests, and we can make the tests by comparing the estimates with their standard errors. We again define the deviance from the largest model considered (the *maximal* model) to be the scale parameter. Type first

```
$FIT HT * GP    $DISPLAY E $
```

giving

```
deviance = 37.575
d.f. = 20

      estimate         s.e.       parameter
1       4.086         11.91       1
2       0.2127        0.06850     HT
3      -3.753         17.53       GP(2)
4       0.009891      0.09997     HT.GP(2)
scale parameter taken as  1.879
```

The interaction is way below its standard error and a parallel line model will be perfectly adequate. Go on by typing

```
$CALCULATE %A = %DV/%DF    $SCALE %A
$FIT - GP.HT    $DISPLAY E $
```

This gives

```
scaled deviance = 20.010
d.f. = 21

      estimate         s.e.       parameter
1       3.279         8.678       1
2       0.2174        0.04989     HT
3      -2.019         0.5846      GP(2)
scale parameter taken as  1.879
```

The GP effect is highly significant. It appears that, while parallel lines fit the data very well, the intercepts are distinctly different. In fact the GP effect, -2.019 ± 0.5846, is exactly the quantity we originally enquired about, the vertical distance between the lines.

What would have been the situation if we had ignored height altogether? Type

```
$SCALE       !FORGET THE PREVIOUS SCALE PARAMETER
$FIT - HT    $DISPLAY E $
```

You will find that the GP effect, which is simply the difference between the sample means of shoulder width in the two groups, is 1.407 ± 0.7555. The height adjustment has had two beneficial effects:

1. It has eliminated the bias due to the two samples having different mean heights.
2. It has allowed us to use the variability about the regression lines in place of the larger variability when height is ignored, giving us a more precise estimate of the quantity of interest.

6.4 Parallel line bioassay

An interesting application of this kind of model is to biological assay data. Suppose a substance produces a response (y) which is linearly related to the logarithm of the dose administered (x):

$$E(y) = \alpha + \beta x$$

say. Suppose too that we have a standard preparation and an unknown preparation which behaves like a dilution of the standard so that a dose d of the unknown produces the same response on average as a dose kd of the standard. The quantity k is known as the *relative potency* of the unknown substance. Taking logs, the equation relating response to log dose for the unknown is

$$\begin{aligned} E(y) &= \alpha + \beta(x + \log k) \\ &= \alpha' + \beta x \end{aligned}$$

where $\alpha' = \alpha + \beta \log k$. Thus we expect the two dose-response lines to be parallel and the log of the relative potency is $(\alpha' - \alpha)/\beta$ which is the horizontal distance between the lines (Fig. 6.4.1).

Table 6.4.1 gives the results of an assay of this kind. The substance concerned is Vitamin D_3 in cod-liver oil. There were three doses of the standard and four of the unknown, and the quantity measured was the percentage of bone-ash from chickens which had the oil added to their diets. Type the data into a file in the layout

```
5.76 33.5 1
5.76 33.0 1
⋮
150 38.4 7
150 40.1 7
```

Table 6.4.1 Bioassay of cod-liver oil for vitamin D_3. Response is % bone ash in chickens.

Dose of standard (BSI units/100g food)				Dose of unknown (mg CLO/100g food)		
5.76	9.60	16.0	32.4	54.0	90.0	150.0
33.5	36.2	41.6	32.0	32.6	35.7	44.0
33.0	36.7	40.5	33.9	36.0	38.9	43.3
32.4	39.5	39.1	31.6	34.8	40.3	44.2
33.7	36.2	39.4	32.7	29.2	42.9	41.8
32.8	35.4	43.0	28.8	34.6	43.9	43.7
37.3	35.6	37.9	30.2	37.7	42.8	38.4
33.1	34.8	42.0	33.1		38.9	40.1
32.1	37.0	42.4			38.6	
29.5	39.4					
	34.2					

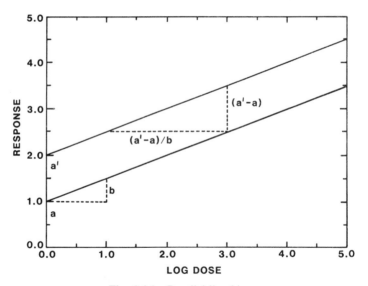

Fig. 6.4.1 Parallel line bioassay.

where the first column is the dose, the second column the response and the third column a code running from 1 to 7 distinguishing between the seven dose groups. Read in the data by typing

```
$UNITS 55    $DATA DOSE Y GP    $DINPUT 8 $
```

As preliminaries we need to take the logs of the doses and also to construct another code PREP equal to 1 for the standard preparation and 2 for the unknown:

```
$CALCULATE LDOS = %LOG(DOSE)
: PREP = (GP>1) + 1       !%GT(GP,1) + 1 IN GLIM3
$FACTOR GP 7 PREP 2 $
```

The first step in the analysis is to get the within-group mean deviance and define this as the scale parameter:

```
$YVARIABLE Y    $FIT GP
$CALCULATE %A = %DV/%DF    $SCALE %A
```

The next question is whether the dose–response curves are reasonably straight. Fit a model with separate straight lines:

```
$FIT LDOS * PREP $
```

The extra scaled deviance is 4.840 on 3 d.f., somewhat above the expected value of 3.0 but not significantly so. A straight-line model is nearly as good as one with an arbitrary pattern of group means.

Next see whether parallel lines will do:

```
$FIT -LDOS.PREP $
```

giving an extra scaled deviance of 0.01945 on 1 d.f., a long way below expectation. A model with parallel lines seems to be a good description of the data.

There is no point in testing for coincidence; it is in the highest degree unlikely that the log relative potency is *exactly* equal to zero, and nobody would be interested if it were. We can go straight on to get the estimates by typing

```
$DISPLAY E $
```

and obtaining

```
      estimate           s.e.        parameter
1       20.02            1.397        1
2        7.394           0.5937       LDOS
3      -14.23            1.336        PREP(2)
scale parameter taken as  4.777
```

From these the log relative potency can be obtained by dividing the difference between the intercepts by the slope, $-14.23/7.394 = -1.925$ (be careful you get the sign correct; for a given dose the unknown gives a *smaller* response, so the relative potency must be less than one and the log relative potency must be negative). Remembering that Glim uses logs to the base e, the relative potency estimate is thus 0.1459. Glim will do the arithmetic for you if you type

```
$CALCULATE %A = -14.23/7.394
: %A   : %EXP(%A)        !NO EQUALS SIGN, SO PRINT THE RESULTS
```

What about a measure of precision for this ratio? Since the slope is many times its standard error we can use the formula

$$\text{Var}(p/q) = (p/q)^2 \{\text{Var}(p)/p^2 + \text{Var}(q)/q^2 - 2\,\text{Cov}(p,q)/pq\}$$

provided we can get the variances and covariances. This you can do by typing

$DISPLAY V $

This gives the triangular table of the variances and covariances of all the estimates:

```
(Co)variances of parameter estimates
1       1.950
2      -0.7907         0.3525
3       1.420          -0.7120         1.786
        1              2               3
scale parameter taken as   4.777
```

The difference in intercepts is parameter 3 and the slope is parameter 2 so the variance of the ratio is

$$(-14.23/7.394)^2 (1.786/(-14.23)^2 + 0.3525/(7.394)^2$$
$$- 2 \times (-0.7120)/(-14.23 \times 7.394))$$
$$= 0.006421 = 0.0801^2$$

You can derive confidence limits for the log relative potency from this figure (using a value of t with 48 d.f.) and by taking antilogs get limits on the relative potency itself.

6.5 The analysis of covariance

So far our models have only contained a single continuous variable and a single factor. Clearly, we can complicate this picture as much as we like. Suppose we have a many-way table with values of a variable y and also of a covariate x. Then we can add a covariate term to the whole set of dummies which define the main effects and interactions of the factors which classify the table, and we can introduce further terms involving interactions (non-parallel slopes) between the several factors and the covariate if the data call for this.

A fairly straightforward example is provided by the data in Table 6.5.1. These relate to a fertilizer trial on rice with four treatments arranged in a Latin square. The main variate y is the yield of rice on each plot. Also available are the yields x on the same plots measured at a time before the different treatments were applied. We are eventually going to adjust the treatment means of y to a constant value of x so as to allow for differences in fertility between the plots which are not picked up by the rows and columns of the Latin square; interpreting the adjusted means can be difficult unless we are assured that the xs cannot themselves have been affected by the treatments. Both the xs and the ys have (for no very good reason) been expressed as percentages of their overall means.

Table 6.5.1 Yields (y) and prior yields (x) on 16 rice plots All yields as % of general mean

	D	B	A	C
x	88	102	91	88
y	90	93	85	81
	B	D	C	A
x	94	110	109	118
y	93	106	114	121
	C	A	B	D
x	109	105	115	94
y	114	106	111	93
	A	C	D	B
x	88	102	91	96
y	92	107	92	102

Input the data by typing along the rows of the table:

```
$UNITS 16     $DATA X Y TR     $READ

 88   90   4
102   93   2
 91   85   1
      . . . .
 91   92   2
 96  102   4

$CALCULATE ROW = %GL(4,4)    :COL = %GL(4,1)
$FACTOR ROW 4   COL 4   TR 4
$YVARIABLE Y $
```

The standard model for a Latin square is one with main effects only and you can fit this by typing

```
$FIT ROW + COL + TR     $DISPLAY E $
```

to produce the estimates

```
          estimate              s.e.          parameter
 1           85.50             8.639          1
 2           21.25             7.727          ROW(2)
 3           18.75             7.727          ROW(3)
 4           11.00             7.727          ROW(4)
 5           5.750             7.727          COL(2)
 6           3.250             7.727          COL(3)
 7           2.000             7.727          COL(4)
 8          -1.250             7.727          TR(2)
 9           3.000             7.727          TR(3)
10          -5.750             7.727          TR(4)
     scale parameter taken as   119.4
```

To add the covariate to the model, simply type

```
$FIT +X    $DISPLAY E $
```

and get

```
          estimate              s.e.          parameter
 1          -16.13            17.31           1
 2           3.247            4.224           ROW(2)
 3           3.070            3.957           ROW(3)
 4           8.677            2.989           ROW(4)
 5          -5.865            3.543           COL(2)
 6          -4.590            3.241           COL(3)
 7          -2.936            3.077           COL(4)
 8          -2.702            2.974           TR(2)
 9           1.258            2.978           TR(3)
10          -0.2330           3.104           TR(4)
11           1.161            0.1942          X
     scale parameter taken as   17.57
```

(Can you see why the estimate of the general mean is so absurd in this analysis?) You will note that the coefficient of x is very highly significant; the residual standard error has come down from $\sqrt{119.4} = 10.93$ to $\sqrt{17.57} = 4.19$, an improvement equivalent to increasing the number of plots almost seven-fold.

The average of the four TR estimates (including the zero 'estimate' for TR(1)) is -0.4192. Subtracting this and adding 100 we can display the adjusted treatment means as

```
   1      2       3       4
100.42  97.72  101.68  100.19
```

Differences between these means will have slightly different standard errors depending on the amounts of adjustment—you can find them as usual by typing $DISPLAY S. For presentation purposes a single approximate standard error found as the average of the six individual errors will usually be adequate. These errors are

1 v 2	2.974
1 v 3	2.978
1 v 4	3.104
2 v 3	2.964
2 v 4	3.185
3 v 4	3.203

an average of ± 3.068 as against ± 7.727 in the original analysis.

7

Counted data

7.1 Counted proportions

So far our y-variables have all been continuous measurements and models assuming a Normal error distribution with constant variance have been reasonable descriptions of the data. Suppose now that the observations we wish to explain or predict are *counted proportions*, so many 'successes' out of so many attempts (we are not talking about measured proportions, such as percentage loss of weight after dieting). An important characteristic of data like these is that the ys—let us change the symbol and call them ps—must lie between 0 and 1. As a result, a straight-line relationship between a p and an explanatory variable will not make sense if taken seriously, since it may predict ps which are greater than 1 or less than 0 (see Fig. 7.1.1). What we need is a curved relationship such as the solid line in Fig. 7.1.1 which flattens out at the top and bottom—an S-shaped curve or *sigmoid* as it is usually called. One such is the *logistic curve* whose equation is

$$p = e^{(a + bx)}/\{1 + e^{(a + bx)}\}$$

A little investigation with a pocket calculator (try it, with $a = 0$, $b = 1$) will show that p tends to 0 and ∞ as x goes from $-\infty$ to $+\infty$ just as we require. The curve is actually symmetric about the $p = \frac{1}{2}$ point which is at $x = -a/b$. The parameter b is a measure of steepness, the slope at the $p = \frac{1}{2}$ point being $b/4$. If we write π for the expected value of p we thus have the model

$$\pi = e^{(a + bx)}/\{1 + e^{(a + bx)}\}$$

To complete the model we must describe the error distribution. The appropriate one for counted proportions is the binomial rather than the Normal. If the expected proportion is not too near 0 or 1, the binomial will be quite well approximated by a Normal distribution but it will not be correct to assume a constant variance. Instead we must use the binomial variance given by $\pi(1 - \pi)/n$ (where n is the number of attempts) and this depends not only upon n but also upon π.

As it stands this appears to be as nasty a non-linear model as you could wish to meet, but it can quite easily be tamed. Replace the proportions p by the *odds* given by $p/(1 - p)$ (so that we talk of 1 *to* 19 rather than of 1 *in* 20) and take logs to the base e. You will find that the model becomes

$$\ln(\pi/(1 - \pi)) = a + bx$$

80

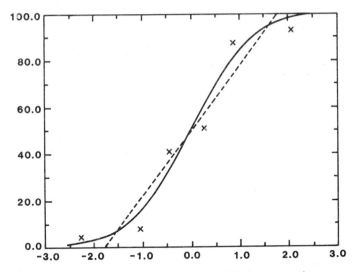

Fig. 7.1.1 Straight line and sigmoid models for proportions

We have a model which consists of a *linear predictor* $a + bx$, but this predicts not $\pi = E(p)$ but a mathematical function of π. This, combined with the binomial error distribution, is a *generalized linear model*, one of the class defined by Nelder and Wedderburn from which Glim takes its name. The function of π which is predicted by the linear predictor is called the *link* function. The particular link which linearizes the logistic curve is called the *logit*; we write

$$z = \text{logit}(p) = \ln\{p/(1 - p)\}$$

7.2 The fitting process

Given some data on counted proportions and an explanatory x-variable, we can calculate the logits of the observed proportions and it is tempting to go ahead and fit a straight line by least squares. There are two snags involved. First of all, the non-constant binomial variance must be allowed for. We can take account of unequal denominators by using weighted regression, but there is also the dependence upon π to be reckoned with. Its effect is that extreme logits, corresponding to πs near 0 or 1, have very much larger variances for a given n and deserve much smaller weights than logits of proportions near 0.5. Secondly, observed proportions equal to 0 or 1 are troublesome. Their logits are infinite, and we cannot trust the symmetric Normal approximation to the binomial at the two extremes of the p-scale.

We shall continue to use the method of maximum likelihood to obtain our estimates, but this will now be more complicated than the least squares methods we have used up to now. With binomial errors we can use a little mathematics to make the arithmetic of maximum likelihood fitting look almost exactly like weighted regression, using a slightly modified z-variable to take care of the ps equal to 0 or 1. Unfortunately, both the weights and the modified working zs depend upon the results of the fitting! The solution to the apparent impasse is to use an iterative method. Start with some kind of approximate fit and derive weights and working zs from this. Then use these to get a new improved fit, re-derive weights and working values, fit again, and continue the process until successive cycles of the iteration agree with each other. Glim takes care of all this complication quite automatically.

Table 7.2.1   Rat fertility assay of Vitamin E

Dose (mg)	Number of rats	Number fertile
3.75	5	0
5.0	10	2
6.25	10	4
7.5	10	8
10.0	11	10
15.0	11	11

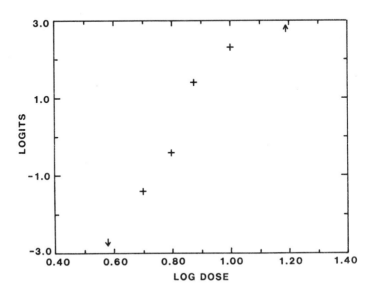

Fig. 7.2.1   Logit (proportion fertile) v log dose of Vitamin E (data from Table 7.2.1)

As an example, some data from a biological assay are shown in Table 7.2.1. Here we have the numbers of fertile rats following treatment with vitamin E at six different doses. Plotting the logits of the observed proportions against log dose (Fig. 7.2.1) gives a rather irregular appearance but shows that a straight line might be an adequate fit. Note the infinite logits denoted by arrows.

We have to tell Glim the doses and the numerators and denominators of the proportions, which we shall denote by R and N. Type

```
$UNITS 6     $DATA DOSE R N     $READ

3.75   0    5       5.0    2    10       6.25    4    10
7.5    8    10      10.0   10   11       15.0    11   11

$CALCULATE LDOS = %LOG(DOSE) $
```

Next we need to specify the error distribution. This needs an $ERROR directive, specifically

```
$ERROR B N $
```

where B stands for Binomial (we shall meet other error distributions later). Notice that at the same time you have to specify the name of the denominator variable.

At this stage it is possible to specify the link function to be used. As we shall see, there are several possibilities within the Glim framework. However, there is a technical sense in which the logit link is especially appropriate to the binomial distribution, and so, if you specify binomial errors Glim will assume that you want a logit link unless you tell it something different.

All that remains is to specify the numerator variable as the thing to be explained and to go ahead with the fit. Typing

```
$YVARIABLE R     $FIT LDOS $
```

produces

```
scaled deviance = 1.1020 at cycle  5
d.f. = 4
```

You may notice that five cycles of iteration have been carried out, and that the deviance has been automatically scaled. With Normal errors you had to estimate the scale parameter from the data and you could specify if you liked that the mean deviance from some model should be used as the scale parameter by way of the $SCALE directive. With binomial errors only the location parameters have to be estimated and we have a theoretical error with effectively infinite d.f. Glim automatically takes this to be the scale parameter unless you provide it with an alternative.

The scaled deviance is equivalent to a suitably weighted sum of squares of deviations from the straight-line model and we can compare it with the theoretical mean scaled deviance of 1.0 as a test of goodness of fit. This simply means interpreting the actual scaled deviance as a value of χ^2 with 4 d.f. Our value of 1.102 is distinctly on the low side, suggesting an exceptionally close fit of the model to the data.

In fact this test needs to be interpreted with caution. The χ^2 distribution is only an approximation to the true distribution of the scaled deviance and theory tells us that the approximation is a good one in large samples. Unfortunately theory does not tell us how large the samples have to be. Practical experience leads to the conclusion that the true significance values of the scaled deviance may often be considerably lower than the numbers in the χ^2 table. The rather low scaled deviance in our example may not be as surprising as it appears to be at first sight.

Let us have a look at the estimates and the fitted values by typing

$DISPLAY ER $

and getting

```
          estimate          s.e.        parameter
1          -11.88            3.298       1
2            6.387           1.746       LDOS
scale parameter taken as   1.000
```

```
unit   observed    out of      fitted    residual
1          0           5        0.155      -0.400
2          2          10        1.674       0.276
3          4          10        4.554      -0.352
4          8          10        7.282       0.510
5         10          11       10.383      -0.502
6         11          11       10.951       0.221
```

The interpretation of the estimates is straightforward. They can be regarded as being at least approximately Normally distributed with the stated standard errors, so that the 95% confidence limits for the slope can be calculated as $6.387 \pm 1.96 \times 1.746$. Note however that this is another of these large-sample results that need a good deal of care, especially when (as here) the samples are anything but large. It will be instructive to remove the LDOS term by typing

$FIT -LDOS $

The change in scaled deviance is 34.747 with 1 d.f. which is to be compared with the square of the value of t obtained by dividing the estimated slope by its standard error—$(6.387/1.746)^2 = 13.381$. With Normal errors and ordinary least squares, these two quantities would have been exactly equal. Here they should be approximately equal but the approximation is clearly pretty poor. It seems likely that the change in scaled deviance is the more to be trusted, but theoretical guidance on this point is scanty.

As to the fitted values, you will see that the observed and fitted numerators are printed. The residuals have been scaled so that they behave roughly like Normal deviates with unit standard deviation.

7.3 Parallel logit lines

Now that we know how to fit a simple regression line to a set of proportions using a logit link, all the other kinds of linear model that we have studied so far become available. Consider as an example the data in Table 7.3.1. These give the numbers of women who had reached the menopause in a set of 2-year age-groups divided into smokers and non-smokers. A plot of the logits (Fig. 7.3.1) of the proportions menopausal in each group suggests that a parallel-line model might be satisfactory and we can investigate this in the usual way (compare Section 6.4).

Table 7.3.1 Numbers of menopausal women by age and smoking habits

Age (years)	Total number	Number menopausal	Total number	Number menopausal
45–	67	1	37	1
47–	44	5	29	5
49–	66	15	36	13
51–	73	33	43	26
53–	52	37	28	25

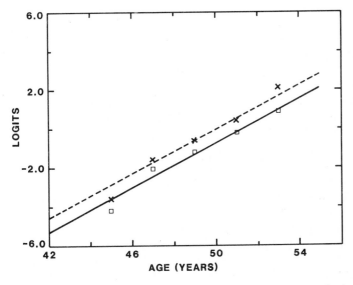

Fig. 7.3.1 Proportions of women menopausal by smoking groups (logit scale)

Type in the data:

```
$UNITS 10    $DATA SMOK AGE R N    $READ
1   5   1   67
1   7   5   44
    . . . .
    . . . .
2  11  26   43
2  13  25   28

$FACTOR SMOK 2    $ERROR B N    $YVARIABLE R $
```

Notice that we have taken an age origin at 40 to make the estimated intercepts a little less silly than usual. Now type

```
$FIT AGE * SMOK
$FIT -AGE.SMOK
DISPLAY E $
```

This produces

```
scaled deviance = 2.2908 at cycle  4
        d.f. = 6

scaled deviance = 2.6368 (change = +0.3460) at cycle  3
        d.f. = 7      (change = +1       )
        estimate       s.e.      parameter
    1    -6.473       0.6083     1
    2     0.7068      0.2462     SMOK(2)
    3     0.5744      0.05647    AGE
    scale parameter taken as   1.000
```

The scaled deviance from the separate lines is a good deal smaller than its degrees of freedom, arguing that the straight lines fit the data well. Removing the interaction term increases the scaled deviance by 0.3460 on 1 d.f., again well below expectation for a 1 d.f. χ^2. The parallel-line model is a good fit and there is a very significant vertical distance between the lines of 0.7068 ± 0.2462 logits. Note in passing that the change in scaled deviance on dropping the SMOK term is 8.3896, to be compared with a squared t-value of $(0.7068/0.2462)^2 = 8.2417$. With the larger samples the approximation of these two to each other is now quite good. As far as interpretation is concerned it may be more meaningful to say that the horizontal distance between the lines is estimated to be $0.7068/0.5744 = 1.23$ years. You can derive a standard error for this ratio just as in Section 6.4, obtaining the variances and covariance by typing $DISPLAY V.

This example gives us an opportunity to describe the logit transform in an interesting statistical way. You may have been struck by the similarity between the S-shaped curve in Fig. 7.1.1 and a cumulative probability curve. In the present example the analogy is particularly close. We can envisage in each of the smoking groups the frequency distributions of exact ages at

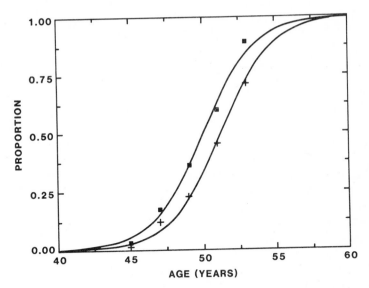

Fig. 7.3.2   Proportions menopausal v age

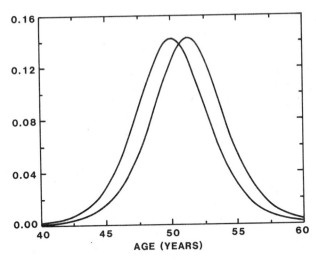

Fig. 7.3.3   Frequency distributions of age at menopause

menopause. These ages are not readily observable in a survey but it is relatively easy to obtain what we have here, sets of independent readings from the cumulative form of the distributions. These readings are the proportions of women whose ages at menopause are less than the given set of ages and they provide the S-shaped curves which have been transformed to straight lines by the logit transform (Fig. 7.3.2). The parallelism of the two lines is equivalent to an assertion that the two distributions are identical apart from a shift along the age axis by an amount estimated at 1.23 years (Fig. 7.3.3).

7.4 Tables of counted proportions

Linear models for logits provide a powerful method of analysis for tabulated data in the form of counted proportions. Here are two examples.

For the first, Table 7.4.1 gives the numbers of cases of a particular disease condition (monoclonal gammopathy, for those who are interested) in the Finistére region of France. The subjects are classified by age and sex and by occupation into farming and non-farming groups. The main point of interest is whether the disease prevalence is different in the two occupational groups. There is an obvious increase in the prevalence rates with age; judging by the bottom marginal totals there appears to be a substantial occupational effect and possibly a sex effect too.

Table 7.4.1 Monoclonal gammopathy in Finistére. (No positive/Total no—rate/1000 in brackets)

Age (years)	Farming M	Farming F	Non-farming M	Non-farming F
30–	1/1590 (0.6)	1/1926 (0.5)	2/1527 (1.3)	0/712 (0.0)
40–	12/2345 (5.1)	7/2677 (2.6)	3/854 (3.5)	0/401 (0.0)
50–	24/2787 (8.6)	15/2902 (5.2)	5/675 (7.4)	4/312 (12.8)
60–	53/2489 (21.3)	38/3145 (12.1)	3/184 (16.3)	1/80 (12.5)
70–	95/2381 (39.9)	63/2918 (21.6)	2/75 (26.7)	0/20 (0.0)
	185/11592 (16.0)	124/13568 (9.1)	15/3315 (4.5)	5/1525 (3.3)

Type the data into a file in the sequence

```
1   1590   1   1926    2   1527   0   712
12  2345   7   2677   ....
```

and set up the problem by typing

```
$UNITS 20    $DATA R N    $DINPUT 8 $
$FACTOR AGE 5    OCC 2   SEX 2
$CALCULATE AGE = %GL(5,4)    : OCC = %GL(2,2)    : SEX = %GL(2,1)
$ERROR B N    $YVARIABLE R $
```

Then the deviances which result from trying a sequence of possible models are shown in Table 7.4.2.

Table 7.4.2 Deviances from various models

	d.f.	Scaled deviance	d.f.	Extra scaled deviance
AGE + OCC + SEX	13	7.6252		
+ AGE.OCC	9	5.7785		
extra for AGE.OCC			4	1.847
− AGE.OCC + AGE.SEX	9	6.7499		
extra for AGE.SEX			4	0.875
− AGE.SEX + OCC.SEX	12	7.2647		
extra for OCC.SEX			1	0.360
AGE + SEX	14	7.9749		
extra for OCC			1	0.350
AGE + OCC	14	34.315		
extra for SEX			1	26.690

None of the two-factor interactions seem to be very important and neither does the main effect of occupation. The main effect of sex is highly significant and that of age is too obvious to need testing. Displaying the estimates for the AGE + SEX model we get

```
      estimate        s.e.       parameter
1       -7.046       0.5004      1
2        1.639       0.5429      AGE(2)
3        2.356       0.5197      AGE(3)
4        3.209       0.5098      AGE(4)
5        3.834       0.5056      AGE(5)
6       -0.5783      0.1142      SEX(2)
scale parameter taken as  1.000
```

If you go on to display the fitted values and residuals, you will find that the fit is good in every cell of the table. It is interesting that the apparent effect of occupation in the bottom margin turns out to be an artefact due to confounding with the other two factors.

Displaying the results of this analysis is not entirely straightforward. It may be enough to set out the fitted values of the prevalence in the two-way age-by-sex table. The fitted numbers are obtainable from the R display and the rates can then be calculated. The results are shown in Table 7.4.3.

The second example comes from a study of snoring as a risk factor for hypertension (high blood pressure) in people over 50 years of age. Table 7.4.4 shows the total numbers and numbers of cases of hypertension classified as positive or negative for snoring, smoking and obesity and also

Table 7.4.3 Fitted values

Fitted rates/1000	
M	F
0.87	0.49
4.46	2.51
9.11	5.13
21.10	11.94
38.72	22.09

Table 7.4.4 Smoking as a risk factor for hypertension. (Number with hypertension/total number)

Smoking	Obesity	Snoring	Men	Women
No	No	No	5/60	10/149
Yes	No	No	2/17	6/16
No	Yes	No	1/12	2/9
Yes	Yes	No	0/0	0/0
No	No	Yes	36/187	28/138
Yes	No	Yes	13/85	4/39
No	Yes	Yes	15/51	11/28
Yes	Yes	Yes	8/23	4/12

by sex. You will notice that there are two empty cells in the table. Glim will not accept zero denominators in a binomial variable so you cannot include these cells and use the %GL function to fill in the factor levels. You must type something like

```
$UNITS 14     $DATA SMOK OBES SNOR SEX R N     $READ
1 1 1 1     5     60
2 1 1 1     2     17
       . . . .
       . . . .
1 2 2 2    11     28
2 2 2 2     4     12
$FACTOR SMOK 2   OBES 2   SNOR 2 SEX 2
$ERROR B N    $YVARIABLE R $
```

An exploration of some possible models gives the deviances shown in Table 7.4.5. There is substantial interaction between smoking and snoring and the obesity main effect is very significant, but sex (rather surprisingly) does not seem to be needed in the model. Displaying the estimates from the fit of SMOK.SNOR + OBES gives

```
      estimate          s.e.        parameter
1       -2.569          0.2489      1
2        0.8244         0.2211      OBES(2)
3        1.129          0.2754      SMOK(1).SNOR(2)
4        1.429          0.4764      SMOK(2).SNOR(1)
5        0.8496         0.3239      SMOK(2).SNOR(2)
scale parameter taken as  1.000
```

Table 7.4.5 Deviances from various models

	d.f.	Scaled deviance	d.f.	Mean scaled deviance
SMOK + OBES + SNOR + SEX	9	15.257		
+ SMOK.OBES	8	15.192		
extra for SMOK.OBES			1	0.065
− SMOK.OBES + SMOK.SNOR	8	5.9399		
extra for SMOK.SNOR			1	9.317
+ SMOK.SEX	7	5.9348		
extra for SMOK.SEX			1	0.005
− SMOK.SEX + OBES.SNOR	7	5.9379		
extra for OBES.SNOR			1	0.002
− OBES.SNOR + OBES.SEX	7	5.4755		
extra for OBES.SEX			1	0.464
− OBES.SEX + SNOR.SEX	7	5.3671		
extra for SNOR.SEX			1	0.573
SMOK.SNOR + OBES + SEX	8	5.9399		
− SEX	9	6.3985		
extra for SEX			1	0.459
− OBES	10	19.529		
extra for OBES			1	13.13

Table 7.4.6 Estimated percentages with high blood pressure

	Non-obese		Obese	
Smoking	No	Yes	No	Yes
Snoring				
No	7.1	24.2	14.9	42.2
Yes	19.2	15.2	35.1	29.0

From these estimates the fitted proportions can be reconstructed as shown in Table 7.4.6. It appears that snoring does increase the risk of hypertension in non-smokers but not in smokers.

In interpreting interactions and additivity in logit models the nature of the logit scale must be borne in mind. If two two-level factors A and B do not interact, the difference between the two levels of B will be the same at both levels of A when measured in logits, but the difference in proportions may differ markedly. A change of 1 logit, for example, takes you from 2% to 5%, from 5% to 12%, from 10% to 23%, from 25% to 48% and from 40% to 64%.

7.5 Binary data

Our counted proportions have all been tabulated for us, either in a one-way table as in the bioassay in Section 7.2 or in more complex ways. What if we are provided with binary (or quantal, or yes/no, or 0/1) data relating to individuals with some continuous covariates describing them? Glim is quite capable of handling data like these, using a binomial error. You simply have to treat each individual as a group of size 1 with either 0 or 1 successes in the group. The parameter estimates will be the maximum likelihood estimates and you can use the changes in deviance as parameters are added or deleted to compare different models. You cannot, however, use the residual deviance to get a measure of goodness of fit of a particular model. The large-sample approximations break down completely when the sample at each set of x-values is only of size 1.

7.6 Alternative link functions for proportions

There is of course no theoretical reason why a logit link should produce a simple linear model for a set of counted proportions, though in practice it often seems to do so. Other link functions are built into the Glim system and can sometimes be useful.

The first of these derives from the interpretation of the logit line as modelling points on a cumulative frequency distribution. The implied distribution is rather unfamiliar and it is natural to ask whether a Normal distribution could be used instead. The corresponding link function is the *Normal equivalent deviate*, also known as the NED or *probit*. The NED of a proportion p is just the point on a Normal curve with mean 0 and standard deviation 1 which has this proportion to the left of it; we have NED(0.5)=0, NED(0.975)=1.96, NED(0.159)=−1.0. (Strictly speaking, the probit of p is the NED plus 5, a measure introduced in the days of desk calculators to avoid the use of negative numbers.) It is not difficult to see that if the cumulative distribution is Normal with mean μ and standard deviation σ, the NEDs of the expected proportions will lie on the straight line $z = \alpha + \beta x$ with $\mu = -\alpha/\beta$ and $\sigma = 1/\beta$.

If you have specified binomial errors you can call for the use of a probit link function by means of the $LINK directive—simply type

$LINK P $

A little experimentation will show you that the fitted values with probit and logit links applied to the same data are almost identical. Very rarely is it necessary (or possible) to choose between the two.

A quite different link function arises in another context. Suppose we have a water sample which may be polluted by bacteria and we wish to know the number of bacteria present. Direct counting under a microscope is possible but laborious and an alternative method is often used. Prepare a number of dilutions of the sample and set up several replicate specimens at each dilution. It is now quite easy to distinguish between those specimens which contain one or more bacteria (in which visible growth will occur) and those which contain none at all.

With good laboratory technique the actual numbers of bacteria in the specimens at a single dilution will follow a Poisson distribution. If the mean is μ the proportion of non-zeros in this distribution is $1 - e^{-\mu}$ and this proportion can be cheaply estimated by counting the number of 'positive' specimens in which growth occurs. Suppose that the number of bacteria in the original sample was N and that the dilution factor was k so that the Poisson means at the successive dilutions were $N, kN, k^2N, \ldots$. Then what we get from our experiment is a set of proportions with expected values

$$\pi = 1 - e^{-N}, \ 1 - e^{-kN}, \ 1 - e^{-k^2N}, \ldots$$

The logs of 1 minus these expected proportions are

$$\ln(1 - \pi) = -N, \ -kN, \ -k^2N, \ldots$$

and if we change the signs and take logs again we get

$$z = \ln(-\ln(1-\pi)) = \ln N, \ \ln N + \ln k, \ \ln N + 2 \ln k, \ldots$$

With the *complementary log–log* link function $z = \ln(-\ln(1-\pi))$ we get a linear predictor of the form $z = \alpha + \beta x$, where $\alpha = \ln N$ (the quantity we are

Table 7.6.1 Dilution series assay

Dilution	No of replicates	No positive
1/1	5	5
1/2	5	5
1/4	5	4
1/8	5	3
1/16	5	2
1/32	5	2
1/64	5	0

interested in), x is the log of the dilution $(0, \ln k, 2 \ln k, ...)$ and the slope is known to be equal to 1.0. This last means that we do not fit the slope but instead include x in the model as an offset (see Section 6.2).

An example is given in Table 7.6.1. The ns are the numbers of replicates and the rs the numbers of positive specimens. The Glim instructions for analysing these data might be

```
$UNITS 7    $DATA DIL N R    $READ

 1 5 5    2 5 5    4 5 4    8 5 3
16 5 2   32 5 2   64 5 0

$CALCULATE LOGK = -%LOG(DIL)
$ERROR B N    $YVARIABLE R
$LINK C       !COMPLEMENTARY LOG-LOG LINK FUNCTION
$OFFSET LOGK
$FIT    $DISPLAY E $
```

This gives

```
scaled deviance = 2.4219 at cycle  4
             d.f. = 6

          estimate          s.e.      parameter
     1       2.063          0.2980       1
     scale parameter taken as  1.000
```

The intercept is the estimated value of z at $\ln k = 0$, i.e. in the undiluted sample. If b is the probability of a positive specimen at this dilution, then

$$\ln(-\ln(1-p)) = 2.063$$
$$\ln(1-p) = -7.870$$

But this is equal to $\ln(e^{-N})$ so that the number of bacteria in the original sample is estimated to be 7.870. 95% confidence limits for this number can be found by taking antilogs of $2.063 \pm 1.96 \times 0.2980$, giving $4.38 - 14.11$. It must be stressed that these limits are based on large-sample theory and may be badly astray with the small numbers we have here.

7.7 Log linear models for contingency tables

Consider next the analysis of a two-way table of counts, a *contingency table* as it is usually called. The ordinary method of analysis for such a table asks whether the row and column factors are *associated* and this question is usually answered by calculating

$$\pm(\text{observed value} - \text{expected value})^2/\text{expected value}$$

the so-called Pearson χ^2. If we denote the marginal total counts in row i and column j by n_{1+} and n_{+j} and the grand total count by n_{++}, then the expected value in the (i, j) cell of the table is $n_{i+}n_{+j}/n_{++}$.

If you look at these expected values you will see that those in the ith row can be written as $v_{ij} = n_{i+} \times (n_{+j}/n_{++})$ This says that the expected proportions in the different columns are the same in each row, and the question 'Is there any association between the row and column factors?' can be rephrased as 'Do the proportions in the different columns vary from one row to another?' From its wording this sounds like a question about interaction.

If we take the log of the expected value we get

$$\ln v_{ij} = -\ln n_{++} + \ln n_{i+} + \ln n_{+j}$$

Written in this form we have a linear predictor for the logs of the expected frequencies—the missing xs can be supplied by creating dummies just as with a continuous y-variable (see Section 4.1). We can thus analyse a contingency table by means of a generalized linear model with a log link function. The error distribution appropriate to the counts is Poisson and you specify this by typing

```
$ERROR P $
```

Glim will supply the log link by default.

An example is shown in Table 7.7.1. The data here relate to 521 10-year

Table 7.7.1   Oral hygiene and type of school

Type of school	Oral hygiene				
	Good	Fair +	Fair −	Bad	
Below average	62	103	57	11	233
Average	50	36	26	7	119
Above average	80	69	18	2	169
	169	208	101	20	521

old children who have been classified by the quality of the school they attend and by the state of their oral hygiene. A set of Glim commands to analyse this table might be

```
$UNITS 12    $DATA N    $READ

62  103   57   11
50   36   26    7
80   69   18    2

$CALCULATE ROW = %GL(3,4)    : COL = %GL(4,1)
$FACTOR ROW 3    COL 4
$ERROR P    $YVARIABLE N
$FIT ROW + COL $
```

which produces

```
scaled deviance = 33.910 at cycle  3
              d.f. =  6
```

As a χ^2 value with 6 d.f. the scaled deviance is highly significant. Not very surprisingly, the state of oral hygiene of a child is quite strongly associated with the quality of the school that she/he attends.

If you calculate the Pearson χ^2 value for this table (you can find this in the system scalar %X2) you will get 31.424, roughly the same as the scaled deviance and with the same degrees of freedom. This approximate equality is once again predicted by theory as a large-sample result. As before, nobody seems to know much about small-sample behaviour, but it seems likely that the scaled deviance will adhere more closely than the Pearson χ^2 to the values in the tables.

You should notice that we have not displayed the estimates of the row and column constants. This is because we are not in the least interested in them. The main effects of rows and columns in this model describe the behaviour of the marginal totals and these are irrelevant to the question about association that we originally posed. Even if there were no interaction, we should not be moved to take out the ROW factor from the model to see whether there were equal numbers in the three school categories. We might call the main-effects model the *minimal* model since we are only interested in the possibilities of departures from it. The model containing only a constant often plays this role, and Glim acknowledges this by automatically including the 'intercept' term in all models unless told not to do so.

We can take the analysis a little bit further. Both the row and column categories are ordered. Suppose we arbitrarily score them 1, 2, 3 and 1, 2, 3, 4. How much of the interaction is explained by the product of these scores? This would be a linear-by-linear interaction. It generalizes the χ^2 test for trend which is often useful for a contingency table with two rows and ordered column categories.

Although ROW and COL have been defined to be factors, we can still do arithmetic on them to produce the product variable. Typing

```
$CALCULATE PROD = ROW * COL
$FIT PROD $
```

gives a scaled deviance of 9.4546 on 5 d.f., with a change of 24.45 for the 1 interaction d.f. The product term mops up a large part of the excess, though the remaining 9.455 (corresponding to $p \cong 0.07$) is enough to suggest that our crude scores are not telling quite the whole of the story.

Once we have defined a log linear model for a two-way table of counts, we can at once extend the notion to more complicated tables. Consider the data in Table 7.7.2. These give some observations on deaths from tetanus in patients treated with and without the use of antitoxin. The patients were divided into groups of more and less severe disease. There appears to be a

Table 7.7.2   Effect of antitoxin on deaths from tetanus

	More severe Antitoxin		Less severe Antitoxin	
	No	Yes	No	Yes
Deaths	22	15	7	5
Survivors	4	6	5	15
Odds (D/S)	5.50	2.50	1.40	0.33

protective effect of the antitoxin; can we say that it differs between the severity groups? What we have here is a structure with three factors each at two levels. The protective effect is described by an interaction between mortality and antitoxin and we want to know if this interaction differs between the severity categories. This is a question about the possibility of a three-factor interaction.

Try typing

```
$UNITS 8   $DATA N   $READ
22  15   7   5
4   6   5  15
$CALCULATE MORT = %GL(2,4)  : SEVE = %GL(2,2)  : ANTI = %GL(2,1)
$FACTOR MORT 2   SEVE 2   ANTI 2
$ERROR P   $YVARIABLE N
$FIT MORT * SEVE * ANTI - MORT.SEVE.ANTI   $DISPLAY E $
```

Notice the shorthand notation for the three-factor model with all the terms fitted except for the three-factor interaction. The display is as follows:

```
scaled deviance = 0.36773 at cycle  4
          d.f. = 1
```

	estimate	s.e.	parameter
1	3.117	0.2064	1
2	-1.884	0.4867	MORT(2)
3	-1.255	0.4060	SEVE(2)
4	-0.4471	0.3198	ANTI(2)
5	1.739	0.5258	MORT(2).SEVE(2)
6	1.096	0.5340	MORT(2).ANTI(2)
7	0.3030	0.5218	SEVE(2).ANTI(2)

```
scale parameter taken as  1.000
```

As far as these rather small numbers are concerned, the scaled deviance indicates that there is little evidence for a three-factor interaction—any protective effect of the antitoxin cannot be shown to differ between the two severity categories. That there may be a protective effect is shown by the MORT(2).ANTI(2) estimate which is just short of twice its standard error. This tells us about its statistical significance, but it would be much more informative if we could interpret the size of the actual estimate, 1.096. This is a linear function of the logs of the expected values, specifically

$$\{(v_{111}-v_{112}-v_{211}+v_{212})+(v_{121}-v_{122}-v_{221}+v_{222})\}/2$$

Taking antilogs, we get the geometric mean of the *odds ratios* in the two groups,

$$\frac{v_{111}}{v_{112}} \Big/ \frac{v_{211}}{v_{212}} \quad \text{and} \quad \frac{v_{121}}{v_{122}} \Big/ \frac{v_{221}}{v_{222}}$$

The antilog of 1.096 is 2.992 so that the antitoxin has multiplied the odds of survival by a factor of around 3.

This example has two special aspects which we ought to consider. First, it has what amounts to a *dependent factor*. We have specified N as our y variable and this is what the model is predicting, but the three factors are not on the same logical footing; we really want to study the effects *of* severity and antitoxin *on* mortality. Moreover, this dependent factor has just two levels. The first consideration means that the minimal model should contain not only all the main effects but also the SEVE.ANTI interaction—we are not at all interested in measuring or testing for any association between the severity and antitoxin factors, which would describe the way in which the antitoxin and no-antitoxin patients were distributed between the severity categories. The second means that we could display the data in another way, in fact as a two-way table of proportions (Table 7.7.3) instead of as a three-way table of counts.

Table 7.7.3 Effect of antitoxin on deaths from tetanus

	Antitoxin	
	No	Yes
More severe	4/26 (15%)	6/21 (29%)
Less severe	5/12 (42%)	15/20 (75%)

If we had started from here we might have thought of fitting a logit model to the proportions. Let us try it now. Type

```
$UNITS 4     $DATA R N     $READ

  4   26      6   21      5   12      15   20

$CALCULATE SEVE = %GL(2,2)     : ANTI = %GL(2,1
$FACTOR SEVE 2   ANTI 2
$ERROR B N    $YVARIABLE R
$FIT ANTI + SEVE    $DISPLAY E $
```

and get

```
scaled deviance = 0.36773 at cycle  4
           d.f. = 1

          estimate          s.e.      parameter
    1      -1.884           0.4868     1
    2       1.739           0.5258     SEVE(2)
    3       1.096           0.5340     ANTI(2)
  scale parameter taken as  1.000
```

You will see that the results, both for the scaled deviance and for the relevant estimates, are exactly the same as before. Reassuringly, under these circumstances the log linear model for the counts and the logit model for the proportions are identical. The logit model is usually a good deal easier to interpret, and the fact that a logit is just the log of the odds explains why an effect in a logit model (which is the difference between two logits) is the log of an odds ratio of some kind.

8

Files and text

8.1 Channel numbers

You have already been using the $DINPUT command specifying the *channel number* of a file from which data are to be read in. Consequently you should know by now how to link a particular named disk file to a known channel number. On some systems you simply have to give the file name in response to a prompt; on others you have to assign the file to the channel at the time the Glim program is loaded, or you may even have to give the file a special name. Once you have mastered this piece of computer technology, there are other ways of using files apart from simply holding data.

8.2 Input channels

So far all the input to Glim (apart from data files read by the $DINPUT command) has been taken from the keyboard. The keyboard is called the *primary input channel* and actually has a channel number of its own—you can find out what this is by typing

```
$ENVIRONMENT C $
```

which on my machine produces the following:

```
Channels:
                use         number   width   height   transcript
    primary     input         1        80
                output        2        80       25
                dump          4
                library       5        80
    current     input         1        80
                echo          *                                  yes
                verify        *                                  no
                warn          2        80                        yes
                fault         2        80                        yes
                help          2        80                        yes
                output        2        80       25               yes
```

This shows that my keyboard is channel 1. We shall meet the other entries in this table later. It is possible to switch to a different source of input by typing

```
$INPUT n
```

where n is a channel number. Glim will be expect the file on this channel to contain commands, possibly with data interspersed, and will proceed to read

100

and obey them.

As an example, take the Bird species dataset that you analysed in Chapter 3. Almost always the analysis of a body of data like this will be an iterative affair. You will want to do some analysis, break off to inspect and brood over the results, return for some more analysis and so on. As the process goes on, you will find that there are some calculation steps that you need to do every time—with the Bird species data, you may decide once for all to transform some of the measurements to logs. You will nearly always want to start with the $UNITS and $DATA commands.

All this suggests that you might construct, not just a data file with numbers in it, but a file with data plus appropriate Glim commands as top and tail to the data. For the Bird species dataset you might construct a file reading

```
$UNITS 43    $DATA SPEC DIST LAT LONG AREA HAB ELEV    $READ

 4  49.9  49.54  6.22   21.4   8   18.0
45   3.2  52.46  4.48  179.8  19  167.0

      .      .      .      .      .      .

16   6.4  54.28  8.40   84.6  11   26.5
19  11.3  55.16  7.12  331.4  12   82.3
$CALCULATE SPEC = %LOG(SPEC)    : DIST = %LOG(DIST)
: AREA = %LOG(AREA)    : HAB = %LOG(HAB)    :ELEV = %LOG(ELEV)
```

If you connect this file to channel 8 and type

$INPUT 8

the data will be read in and the calculations performed.

A few things are still missing from the file, though. First it is very desirable to put a comment (starting with $C) at the head of the file to give some identification details. Nothing is more infuriating than a disk file obviously containing Glim material when you cannot remember just what material it is. Next you may find it worrying that nothing appears on the screen as the file is read in. To cause the input lines from the file to be displayed type

$ECHO

right at the front of the file. You presumably will not want the keyboard input echoed when you come back from reading the file, since every line will then appear twice. Accordingly you should insert another $ECHO directive near the end of the file. The $ECHO directive is an example of a *toggle* which affects the state of the Glim program in some respect, simply reversing the state from off to on or from on to off.

Finally and most importantly, you cannot just let Glim run off the end of the file. The very last command to appear should be

$RETURN

which will take you smoothly back to the primary input channel.

Used as described, channel 8 would be called the *secondary input channel*. There is no reason why the commands read from this channel should not include another $INPUT command, say to channel 9. Provided that this channel is connected to a file, input will be taken from it so that it could be called the tertiary input channel. When the $RETURN directive is encountered on channel 9, input will be switched back to the secondary channel 8.

8.3 Output channels

Just as input is normally from the keyboard (the primary input channel), so output, at least when Glim is used interactively, is usually to the screen (the *primary output channel*). You can divert output from the screen to a file connected to a channel by typing

$OUTPUT n $

where n is the channel number. You will have to resign yourself to not seeing the output material until you can examine the file later on. On most systems it is possible for the channel to be a printer, so that this is one of the ways of getting hard copy. It is often possible, though, to get everything that appears on the screen to go automatically to a printer or to a file (in Glim77 such a file is called the *transcript file* and must be called for at the start of the session).

If you type the $OUTPUT directive with no channel number, output will be switched off altogether. This is occasionally useful in advanced uses of Glim when you are doing a long iterative calculation of which only the conclusion is of any real interest.

8.4 Widths and heights

You can specify the *width* of an input or output channel (i.e. the number of characters per line) by way of a number following the channel number in the $INPUT or $OUTPUT directive. For input the width must be between 30 and 299; for output it must be between 52 and 132. Restricting the input width can be useful in avoiding the reading of identification material typed at the right-hand ends of the lines. Changing the output width also changes the widths of plots (Section 1.7)

A second additional number in an $OUTPUT command controls the value of the *height* of the output channel, the number of lines on a 'page'. This too affects the size of plots. The height must not be less than 6.

@ @ @ @
In Glim77 the directive

$PAGE

causes output to the screen to stop at the end of each page. This directive is a toggle, so repeating it turns the facility off again.

@ @ @ @

Of course the width specified for a channel must be consistent with any limitations imposed by the hardware or by your local system. You will have to investigate this yourself.

8.5 Dumping and restoring

If you want to break off in the course of a Glim analysis, either to pause for thought or to answer the telephone, you will probably want to keep the variables and other material that you were using at the time. You can do this by typing

$DUMP $

There is a channel number allotted to dumps (typing $ENVIRONMENT C will tell you what it is) and I am assuming that this channel is connected to a suitable file. To get back the same material at a subsequent session type

$RESTORE $

The $DUMP command leaves the file at a point following the last item dumped. This means that you can make several dumps in the course of a single Glim session and they will be filed one after the other. If in a subsequent session you type $RESTORE more than once, the last dump restored will overwrite the earlier ones. Thus three successive $RESTOREs will bring back the third of a sequence of dumps. However, you can issue the command

$REWIND $

to get back to the beginning of the dump file. You will need this if you want to restore a dump made earlier in the same session.

The $REWIND command will work on a channel other than the dump channel—this is useful since the $READ and $DINPUT commands leave their files at the point following the last item read. All you need do is to give the channel number immediately following the $REWIND keyword. You can in fact do the same with the $DUMP and $RESTORE commands if you want to dump to a channel other than the pre-set one.

8.6 Printing and text

Usually Glim will output automatically all you need to know about its results. However, it is sometimes useful to add to the output and the $PRINT command enables you to do this. This directive can be followed by any number of vector and scalar names, and the corresponding items will then be displayed. They will start on a new line and then appear one after another across the lines according to the width of the current output channel. Try setting the width of your output channel to 60 (you can find out how to do this in Section 8.4; the channel number and its current width are part of the $ENVIRONMENT C display). Then type

```
$UNITS 12
$CALCULATE A = %GL( 4 , 3 )     : %A = %CU( A )
$PRINT A    %A    %NU
```

(Remember that %NU is the system scalar equal to the current number of units.) The result will be

1.000	1.000	1.000	2.000	2.000	2.000	3.000
3.000	3.000	4.000	4.000	4.000	30.00	12.00

For labelling purposes you will probably want to intersperse text among your printed items. This is quite easily done by including the text in the proper place in the list of items to be printed, surrounded by single quotes. Thus you might follow a $FIT command by

```
$CALCULATE %M = %DV/%DF
$PRINT
: 'MEAN DEVIANCE =' %M $
```

Because each $PRINT command starts a new line, the first $PRINT command with no items following it will produce a blank line.

@@@@
In Glim77 you can start a new line in the course of a $PRINT command by putting a semicolon into the print list at the required place.
@@@@

The number of digits printed in a number is normally that given by the current setting of the $ACCURACY directive (Section 1.10). More digits will be printed if the integer part of the number needs them. You can also specify different accuracies for individual items in the print list by preceding them by *n, where n is the accuracy required. If n is negative, following items will be printed rounded to the nearest whole number; if n is zero, following items will be ignored. A setting of n stays in force until another *n is encountered or until the end of the $PRINT command. You should experiment with this to see how it works. Type

```
$UNITS 8
$CALCULATE A = %CU( 1 ) - 3     : B = 10 ** A $
```

and then issue some commands to print B with different accuracies.

@@@@

Quite apart from the $TPRINT command (Section 5.2), Glim77 has many more possibilities in the $PRINT command which give you better control over the printed layout. As a simple example

```
$PRINT *REAL %A 8,3    *INTEGER %B 4 $
```

would print %A in a field of width 8 characters with 3 decimal places and %B as an integer in a field of width 4 characters. Full details are in Section 5.2 of the *User's Manual.*

@@@@

9

Macros

9.1 What is a macro?

I hope you have already been convinced that Glim is a very powerful tool for fitting linear models and their generalizations to all kinds of data. Its scope can however be made even wider than this. This is because you can yourself define new Glim instructions (made up of combinations of existing ones) give them names, store them and subsequently use them. Such a super instruction is one form of Glim *macro*.

Essentially the idea of a macro is extremely simple. A macro is nothing more than a string of characters, usually making up a set of Glim commands, which is stored by the program just like the values of a variable and which can be used as input to the program rather like the contents of disk file on a subsidiary input channel. The notion and its uses can best be explained by some examples.

9.2 Simple macros, substitution

Suppose we are fitting a logit or log linear model and that Glim has evaluated the scaled deviance and its degrees of freedom. Treating it as a χ value, what is its significance probability? You could look it up in tables, but you can get a good approximation by writing x for the χ^2 value and d for the d.f. and calculating

(1) for $d=1$, $(1-NP\{\sqrt{x}\}) \times 2$,
(2) for $d=2$, $\exp(-x/2)$,
(3) for $d>2$, $1-NP[\{(x/d)^{1/3} - 1 + 2/(9d)\}/\sqrt{\{2/(9d)\}}]$,

where NP stands for the left-hand tail area in a Normal distribution with mean 0 and standard deviation 1. It would be intolerable if you had to type out the equivalent string of Glim instructions every time you needed to do the calculation. Instead let us define them to be a macro whose name is CHP1. Type

```
$MACRO CHP1
$CALCULATE %P = %EQ(%DF,1) * (1 - %NP(%SQR(%DV)))
    + %EQ(%DF,2) * (%EXP(-%DV/2))
    + %GT(%DF,2) * (1 - %NP(((%DV/%DF) ** 1/3 - 1 + 2 / (9 * %DF
    / %SQR(2/(9 * %DF))))
$
$ENDMAC
```

(Note the use of the relation functions %EQ, %GT to pick out the required

106

:ase. In Glim77 these could have been written $(\%\text{DF}==1)$, $(\%\text{DF}==2)$, $(\%\text{DF}>2))$. Now if you follow a \$FIT command by a command to use the macro—

\$USE CHP1   \$PRINT %P \$

the whole complicated instruction will be obeyed and the probability will be calculated and printed out. You can try this by putting some numbers into %DV and %DF using \$CALCULATE commands and comparing the results with what you find in the tables.

One caution—when a macro consists of one or more Glim commands, it is very important to terminate it by a \$ sign over and above that belonging to the final \$ENDMAC directive. This latter does not belong to the macro, and without the final \$ sign the last command in the macro may not be obeyed properly.

The calculations of CHP1 can be specified in a slightly different way. CHP1 uses the scalar %P and this may be a nuisance if %P is already being used by the surrounding Glim program in another context. Define a second macro CHP2 which is exactly the same as CHP1 leaving out the piece

\$CALCULATE %P =

at the beginning. This new macro is still a string of characters although it is no longer a complete Glim instruction. You can arrange for this string to be read in and inserted into the Glim input by means of the *substitution symbol*, usually #. Instead of the \$USE command you could then type

\$CALCULATE %P = #CHP2 \$

with exactly the same effect as before. If you only wanted the probability to be printed without being stored,

\$CALCULATE #CHP2 \$

would be enough, there being now no = sign in the \$CALCULATE command. Notice that you could have typed

#CHP1

in place of the \$USE command.

9.3 Macros with arguments

The macros CHP1 and CHP2 are potentially quite useful but the time will undoubtedly come when you find that you want the significance probability of a χ^2 value which is not the scaled deviance following a \$FIT command. You could of course put the value and its degrees of freedom into %DV and %DF using \$CALCULATE commands before using the macro, but this is a bit clumsy. Instead, define a third version of the macro and call it CHP3. This should be exactly the same as CHP1 except that %DV should be replaced by %1 wherever it occurs, and %DF by %2. The apparent scalars %1 and %2 are actually *formal arguments* of the macro (the idea will be familiar to any programmer who has written or used a subroutine or procedure with arguments). Before you use CHP3 you will have to specify the *actual arguments* which are to be substituted for the formal ones on this particular occasion. This needs an $ARGUMENT directive. If you type

```
$ARGUMENT %DV %DF    $USE CHP3
```

this will have exactly the same effect as using CHP1. When CHP3 is obeyed any occurrence of %1 will be replaced by the first actual argument %DV and any occurrence of %2 by the second actual argument %DF. Note that an assignment of actual arguments by an $ARGUMENT directive persists, perhaps through several $USEs, until it is overridden by another one. Actual arguments in a subsequent $ARGUMENT directive that are not to be reset can be denoted by *s, or simply omitted if they come at the end of the argument list.

@@@@

In Glim77 the list of arguments can follow the macro name in the $USE directive instead of being in a separate $ARGUMENT directive. Some confusion can arise if the $USE and # methods of calling for macros are combined in the same program—Section 11.4.1 of the *User's Guide* explains why.

@@@@

Actual arguments must be names, not constants or expressions, since they are effectively copied into the appropriate places in the macro text when it is used. However, macros can contain $USE commands for other macros (or even for themselves) and actual arguments can be the arguments of enclosing macros. Thus in a argument list %2 means 'the second actual argument of the enclosing macro (or main program)'. If you like puzzles you might try working out what the following nest of macros does:

```
$MACRO M    $CALCULATE %A = %1 $    $ENDMAC
$MACRO N1    $USE N2 $    $ENDMAC
$MACRO N2    $USE M $    $ENDMAC
$ARGUMENT N1  %B
$ARGUMENT N2  %C  %1
$ARGUMENT M  %2
$USE N1 $
```

Section 16.4 of the *Glim3 Manual* (Section 11.4 of the Glim77 *User's Guide*) will give you the answer.

@@@@

Glim77 provides a set of scalars %Z1, ..., %Z9. These are available for general use but it is recommended that they be only used inside macros so as

to avoid clashes with the scalars %A to %Z which may be used by the surrounding program. Glim77 also allows names to include the underline symbol, so that A_3 and CDE_ are legitimate names. It is recommended that names which end with the underline symbol be reserved for working variables inside macros for the same reason.
@@@@

9.4 Programming with macros

To turn Glim into a real programming language, two facilities are needed, branching and looping. The first of these is provided by a sort of 'case' construction using the $SWITCH command. This directive is followed by a scalar name and a list of macro names. If the scalar has the value n, the nth macro in the list (if this exists) is called for; if n is less than 1 or greater than the number of macros in the list, the command has no effect.

More important is the looping command $WHILE. If you type

```
$WHILE scalar name    macro name    $
```

the macro will be used repeatedly (with its current arguments if it has any) as long as the scalar is 'true', i.e. non-zero. Presumably the scalar's value will be changed inside the macro; otherwise the macro will be used either indefinitely or not at all.

This command is well suited for programming iterative approximation routines. Suppose we make a macro which, when used with a scalar argument x, puts the value of some function $f(x)$ into %F. Suppose too that we know that $f(x)$ has a root between 0 and 1 and that the function is increasing over this range. Then we can calculate the root by bisection something as follows:

set $A = 0$, $B = 1$
calculate $x = (A + B)/2$, $F = f(x)$
if $F < 0$ set $A = x$; else set $B = x$
repeat while $(B - A)$ is greater than a tolerance

Translating this into Glim needs a macro to set up the iteration and another for the loop. We shall use the first argument to denote the function (which will have to be calculated by yet another macro) and the second argument for the tolerance. Try typing

```
$MACRO SOL1
$CALCULATE %A = 0   : %B = 1   : %X = 0.5   : %D = 1 !I.E. TRUE
$ARGUMENT SOL2 %1 %2   !TRANSFER THE ARGUMENTS TO SOL2
$WHILE %D SOL2   !LOOP UNTIL %D GOES FALSE
$
$ENDMAC

$MACRO SOL2
$USE %1  !CALCULATE F(X)
$CALCULATE %C = (%F < 0)        !%LT(%F,0) IN GLIM3
:          %A = %IF(%C,%X,%A)
```

```
:               %B = %IF(%C,%B,%X)
:               %D = (%B - %A) >= %2    !%GE(%B - %A, %2) IN GLIM3
:               %X = (%A + %B )/2
$
$ENDMAC
```

You could try this on the function $f(x) = e^x - 2$. You will need to define

```
$MACRO FUNC
$CALCULATE %F = %EXP(%X) - 2
$
$ENDMAC
```

and then type

```
$CALCULATE %Z = 0.005      !THE TOLERANCE
$ARGUMENT SOL1 FUNC %Z
$USE SOL1
$CALCULATE %X      !NO EQUALS SIGN - PRINT THE RESULT
:              %LOG(2)     !AND CHECK IT
```

You may want to put a command into the inner macro to print the values of
%A and %B to track the progress of the iteration.

9.5 Macros in files and subfiles

Glim macros provide a means by which sets of Glim commands for doing
special jobs can be put on record and communicated from one user to
another. The *Glim Newsletter* regularly contains listings of macros of general
interest. To make these and others available when you need them, you will
naturally type them into files; but it seems extravagant to use one file per
macro, and indeed this may be beyond the scope of your local system.
Instead you can arrange for each macro (or set of related macros like SOL1
and SOL2 in the previous section) to be stored as a *subfile* in a single file.

A subfile is a block of material (in this case a set of Glim commands
making up a macro) starting with the directive

$SUBFILE subfile name $

which (exceptionally) *must* start at the beginning of a line. A subfile is
terminated by a $RETURN directive. If a file consists of a set of subfiles, its
final statement must be the end-of-file directive

$FINISH

This too *must* start at the beginning of a line. To input a subfile, simply place
the subfile name following the channel number in the $INPUT command.

An $INPUT command for a whole file leaves the file at its end, so that in the
absence of a $REWIND command a subsequent $INPUT command to the
same channel will fail. A subfile on the other hand will be looked for from
wherever the file was last left, going back to the beginning if necessary. In
either case you can start reading from the beginning of the file by replacing

he $INPUT directive by $REINPUT.

@@@@

    Glim77 comes with a whole library of general-purpose macros held as ubfiles on an input channel whole number is kept in a system scalar %PLC. There is an information subfile called INFO, so you should type

$INPUT %PLC   80   INFO $

with echoing (Section 8.2) switched on to see what you have got. More detailed information should be available on a separate text file, but you will have to investigate this for yourself.

@@@@

.6 Macros and subsidiary input

The effect of calling for the use of a macro is to take the Glim program from he stored material rather than from the keyboard. This is very reminiscent of input from a subsidiary input channel, and some facilities are common to both forms of input. In particular, the subfile mechanism is quite general and subfiles to be read by an $INPUT command can contain Glim commands and/or data as required.

    Both inside a macro and during input from a subsidiary channel, it may be handy to return temporarily to keyboard input so that the user can control what is going on to some extent. This can be done by including the command

$SUSPEND

in the file or macro. When you eventually type

$RETURN

input will go back to the channel or macro that it came from.

    There is a mechanism for returning from a macro or subsidiary input channel to a higher program level from which the macro or file was entered. This can sometimes be useful when the Glim program detects an error. The necessary command is

$EXIT n $

If $n = 1$, exit is to the 'immediately surrounding' level, i.e. to the point following the preceding $INPUT, $USE, $WHILE, $SWITCH command. If $n = 2$, the exit is to the level surrounding this, and so on. If a macro is used via a $WHILE command, an alternative is

$SKIP n $

which jumps to the end of the loop at the appropriate level.

    It is quite legitimate for macros to contain $USE, $SWITCH and $WHILE commands calling upon further macros, including themselves. They can also contain $INPUT commands, and $USE, $SWITCH and $WHILE commands can be read from subsidiary input channels. This is fairly straightforward if all goes well, but reckoning out the result of an $EXIT or $SKIP command can get very complicated. If you can understand Section 16.3 of the *Glim3 Manual*, or Section 11.6 of the Glim77 *User's Guide*, you should have no further problems.

@@@@

9.7 Echoing macro contents

You may have discovered by now that the instructions in a macro are not normally echoed to the screen as they are obeyed. In Glim77, echoing of macro contents can be switched on and off by the toggle command

`$VERIFY`

@@@@

10

Miscellanea

This chapter contains some sections which do not fit conveniently elsewhere in the book.

10.1 Convergence of an iterative fit

In Chapter 7 we cheerfully stated that iterative cycles went on occurring until convergence took place. There are situations in which the iterative process cannot be expected to converge since the best estimates are infinite. This will happen, for example, if you try to fit a logit line to a set of proportions 0, 0, 0, 1, 1, 1 with $x = 1, 2, 3, 4, 5, 6$. Iteration will continue for 10 cycles and then a message

NO CONVERGENCE BY CYCLE 10

will be displayed. A display of the estimates will disclose a very large value for the slope, which should actually be vertical. Other rather more subtle situations of the same kind can also occur when you fit log linear models to tables with systematic patterns of zero counts.

If during iteration the deviance actually increases sharply, the process stops with the message

DIVERGENCE AT CYCLE n

This almost never occurs in practice.

Other problems can happen with links like the logit and log which can give infinite values for finite observations. The message

UNIT n HELD AT LIMIT

means that the linear predictor was tending to ∞ and that the fitted proportions have been fixed at 0% or 100%. It may mean that an unsuitable model was being used.

You may also get the message

CHANGE IN D.F.

This happens when the variance of a parameter becomes so large that it appears to be aliased (see Section 4.5), i.e. wholly confounded with the other parameters in the model. If iteration continues, a larger deviance may 'unalias' the parameter and the deviance will proceed to oscillate. Again the model being fitted may be inappropriate.

The limit of 10 on the iteration cycles permitted is a default setting and can be changed by the $CYCLE directive. This takes the form

$CYCLE   no of cycles   display frequency $

so that you can specify the number of cycles permitted and how often (once per three cycles, say) the deviance is to be displayed. If the second number is zero or omitted, the display is made only at the end of the iteration.

There is also a similar $RECYCLE directive, to be used with or without parameters. It specifies that iteration shall use as starting values the fitted values from a previous $FIT. If the effect of adding or deleting x-variables is expected to be slight, these old fitted values may be better starting points than the usual ones obtained by Glim *ab initio*.

@@@@

Glim77 allows you to set the actual tolerances used to decide convergence and aliasing by including one or two further parameters in these directives. The default values are 10^{-4}.

@@@@

10.2 Survival data

Suppose that the quantity we wish to explain or model is a *survival time*, the time from an origin (such as the application of an anti-cancer treatment or the placing of a machine part on trial) to the occurrence of some event (the patient dies or goes into remission, the machine part fails). Data like these have two characteristics. The distribution of survival times is usually very skew (an exponential distribution with its mode at the origin is sometimes met with), and the data are often subject to *censoring*. This means that, for some of the units, the event has not occurred by the end of the observational period so that it is only known that their survival times are greater than a certain amount. Maximum likelihood is possible assuming exponential or more complicated forms for the survival time distribution, and D. R. Cox in a classic paper (Regression models and life tables, *J. roy. statist. Soc.* B, 34, 187–220, 1972) showed how modelling techniques could be used without making any assumptions about this distribution.

Referring to the event for convenience as death, suppose the fraction surviving longer than a time t is $S(t)$, so that $f(t) = -dS(t)/dt$ is the survival time distribution. Then the ratio $f(t)/S(t)$ is called the *hazard*—it is the probability of dying around time t given that the unit has survived up to that time. Cox's model assumes that the hazard function for a particular unit can be written in the form

$$h(t) = \lambda(t) \exp(\eta)$$

where η is a linear predictor and the function $\lambda(t)$ is arbitrary but common to all the units. This means that the ratio of the hazards of two units is constant over time and the model is called one of *proportional hazards*.

M. Aitkin and D. Clayton (The fitting of exponential, Weibull and extreme-value distributions to complex censored survival data using GLIM. *Appl. Statist.*, 29, 156–63, 1980) showed that Glim could be used to fit models with certain known survival time distributions, and J. Whitehead (Fitting Cox's regression model to survival data using GLIM. *Appl. Statist.*, 29, 268–75, 1980) extended their methods to deal with the Cox model. Applied to the problem of comparing two groups, Whitehead's method involves presenting, for each time at which a death occurs, the data on N, the number at risk of dying at that time (i.e. the number that have not already died or been censored), and m, the number of deaths in each group. The model now states that the hazards in the two groups at time t are $\lambda(t)$ and $\lambda(t)$ $\exp(\beta)$. It turns out rather surprisingly that we can fit this model by taking the ms to be the y-variable, specifying Poisson errors and a log link, and fitting two factors, one with two levels for the two groups and the other with one level for each of the distinct times. It is also necessary to include $\ln N$ in the model as an offset. A slightly different method involving binomial errors and a logit link is described by C. Jagger and D. Clayton in Issue 5 of the Glim Newsletter (December 1981) where a comprehensive Glim program is given.

10.3 Gamma errors

An alternative error distribution provided by Glim for continuous y-variables is the gamma distribution (a multiple of a χ^2 distribution). This may be useful for a y which is positive and has a constant coefficient of variation rather than a constant variance.

A class of models for which this is appropriate is that of the *inverse polynomials* introduced by J. A. Nelder (Inverse ploynomials, a useful group of multi-factor response functions. *Biometrics*, 22, 128–141, 1966). In general for a single x-variable these take the form

$$x/y = a + bx + cx^2 + \cdots$$

so that we have a linear predictor for x/y and what amounts to a reciprocal link. This is the natural link for the gamma error distribution and Glim will provide it by default. The inverse linear curve is

$$1/y = a + b/x$$

while the inverse quadratic is

$$1/y = ax + b + c/x$$

where all the coefficients are positive. The first of these rises to an asymptote at $y = 1/a$. The second rises to a maximum and then falls again but is not symmetric about the maximum like an ordinary parabola. The equations can easily be generalized to include powers and products of two or more different *x*s on the right-hand side.

A constant proportional error may well be appropriate for an essentially positive variable and we could fit one of these curves using the gamma error distribution. Let us try fitting an inverse linear relationship to the following data:

```
x  2     4     6     8     10
y  6.83  7.99  7.79  9.45  8.19
```

Type

```
UNITS 5       $DATA X Y     $READ

2 6.83    4 7.99    6 7.79    8 9.45    10 8.19

$ERROR G     !SPECIFY GAMMA ERRORS
$YVARIABLE Y
$CALCULATE XX = 1/X
$FIT XX      $DISPLAY E $
```

and get

```
deviance = 0.018237 at cycle   3
    d.f. = 3

          estimate         s.e.       parameter
   1        0.1075      0.008036      1
   2        0.07748     0.03296       XX
   scale parameter taken as  0.006079
```

A plot of 1/Y and 1/%FV against 1/X (try it) shows that the fit is about as good as could be expected with these rather irregular data.

Another application of the gamma distribution is to the analysis of variance of a balanced table with three or more random factors. Under these circumstances a test of significance for a main effect is tricky since there is no single mean square with the correct expectation to act as an error term. Some *ad hoc* recipes are available for constructing combinations of mean squares to use for error in what are called pseudo-*F* tests (see for example H. Scheffe, *The analysis of variance*, Wiley 1959, Section 7.5). Linear modelling provides an alternative. It is easy to write down the expectations of the mean squares in terms of the variance components, and assuming Normality the observed mean squares will be independent gamma variables. We can thus set up a generalized linear model in which the variance components are the unknowns and the observed mean squares the *y*-variable values. The degrees of freedom are used as weights and this provides a theoretical error with a value of 2.000.

Consider the analysis of variance table (Table 10.3.1) relating to an experiment with three factors at 10, 6, and 4 levels. The coefficients of the

Table 10.3.1 Analysis of variance

	d.f.	MS
A	9	147.86
B	5	48.11
C	3	116.57
AB	45	38.22
AC	27	19.15
BC	15	27.62
ABC	135	12.43

Table 10.3.2 Coefficients of variance components

Mean square	σ_A^2	σ_B^2	σ_C^2	σ_{AB}^2	σ_{AC}^2	σ_{BC}^2	σ_{ABC}^2
A	24	0	0	4	6	0	1
B	0	40	0	4	0	10	1
C	0	0	60	0	6	10	1
AB	0	0	0	4	0	0	1
AC	0	0	0	0	6	0	1
BC	0	0	0	0	0	10	1
ABC	0	0	0	0	0	0	1

variance components in the expectations of these mean squares are shown in Table 10.3.2. To analyse these data you need to specify a gamma error distribution and an *identity* link since it is the y-variable itself which is to have a linear predictor. Type

```
$UNITS 7     $DATA DF MS A B C AB AC BC ABC     $READ

  9 147.86 24  0  0  4  6  0  1
  5  48.11  0 40  0  4  0 10  1
135  12.43  0  0  0  0  0  0  1

$YVARIABLE MS
$ERROR G     $LINK I  !SPECIFY IDENTITY LINK
$WEIGHT DF       $SCALE 2.0
$FIT A + B + C + AB + AC + AC + ABC - %GM     $DISPLAY E  $
```

This gives a perfect fit and the resulting estimates are simply those obtained in the usual way:

```
scaled deviance = 0. at cycle   1
              d.f. = 0
```

```
            estimate            s.e.        parameter
1             4.288           2.932         A
2           -0.1325           0.8272        B
3             1.370           1.598         C
4             6.448           2.050         AB
5             1.120           0.9045        AC
6             1.519           1.020         BC
7            12.43            1.513         ABC
scale parameter taken as   2.000
```

The estimated variance component for *B* is negative. This may be taken as an indication that the true value is small and we could try removing it from the model. Typing

$FIT - B $DISPLAY E

produces

```
scaled deviance = 0.023171 (change = +0.02317) at cycle  3
          d.f. = 1         (change = +1        )
            estimate        s.e.       parameter
1             4.300         2.931       A
2             1.377         1.597       C
3             6.379         1.982       AB
4             1.118         0.9045      AC
5             1.477         0.9645      BC
6            12.44          1.513       ABC
scale parameter taken as   2.000
```

As usual the standard errors are large sample approximations as is the Normality of the estimates, and caution is needed in the interpretation. Notice, though, that the change in scaled deviance on removing B is not too far from the (estimate/s.e.)^2 of the preceding fit.

10.4 Space control

Glim of course has only a limited amount of space within the computer store at its disposal and it has to hold there all the variable values plus macro texts and some behind-the-scenes material of its own. You can get an idea of how you are doing for space at any time by typing

$ENVIRONMENT U $

If you run short of space you may want to get rid of some of the variables or macros that you reckon you can do without. This can be done with the $DELETE command. Type

$DELETE   list of names $

where the names are those of the variables and macros you want removed.

A few tricks can be used to ensure that a macro takes up as little space as possible. You should abbreviate all the directives, at least down to four characters and even further if you can recall or look up the permitted minimum abbreviations. Put an end-of-record symbol (!) at the right-hand end of each line, otherwise a lot of blanks will be stored. Using the repetition symbol (:) as much as possible not only saves space but also saves time since the directive name does not have to be re-decoded. As an example, the macro SOL2 in Section 9.4 could be stored as

```
$M SOL2!
$U %1!
$CA %C=%LT(%F,0):%A=%IF(%C,%X,%A):%B=%IF(%C,%B,%X)!
:%D=%GE(%B-%A,%2):%X=(%A+%B)/2!
$$END
```

10.5 System scalars

Here is a list of the more useful system scalars.

%DV, %DF   Equal to the (scaled) deviance and its degrees of freedom following a $FIT.

%ML   Equal to the length of the %VC vector (see the next section).

%NU   Equal to the number of units from the $UNITS directive.

%PI   Set initially to 3.14159265...

%PL   Equal to the number of non-aliased parameters, the length of the %PE vector (see the next section).

%SC   Equal to the scale parameter if set; else to the mean deviance following a $FIT.

%X2   Equal to the Pearson chi-squared (or a generalization of it) following a $FIT.

@@@@
    Glim77 has a much larger number of system scalars. These include

%A1 to %A9   If the ith parameter of a macro is set, %Ai is 'true' (equal to 1) else it is 'false' (equal to 0).

%Z1 to %Z9   Generally available but intended to be used as temporary variables in library macros.

%ACC   Equal to the $ACCURACY setting (see Section 1.10)

%PIC, $PIL,

%CIC, %CIL   Equal to the channel numbers and widths of the primary and current input channels

%POC, %POL,

%POH, %COC,

%COL, %COH   Equal to the channel numbers, widths and heights of the primary and current output channels.

%PDC, %PLC   Equal to the channel numbers of the dump channel and the macro library.

For the others see Section 2.2.2 of the *Reference Guide*.
@@@@

Note that the values in the system scalars are copies of those used by the system, so that you cannot change the latter by assigning values to the former. A command

$CALCULATE %NU = 10 $

is *not* equivalent to

$UNITS 10 $

10.6 System vectors

Here is a list of some of the system vectors—

%BD   Contains the binomial denominators after $ERROR B.
%FV   Contains the fitted values after a $FIT.
%LP   Contains the linear predictors after a $FIT.
%OS   Contains the offset values after an $OFFSET directive.
%PW   Contains the prior weights after a $WEIGHT directive.
%WT   Contains the iterative weights after a $FIT.
%YV   Contains the y-values after a $YVARIABLE directive.

All these have the standard length given by the $UNITS directive. Note that %BD, %OS, %PW and %YV are actually *pointers*; they are, as it were, synonyms for the appropriate user vectors.

If the system vector %RE has values, then units whose value in %RE is zero are ignored by a $PLOT command, and also by a $DISPLAY W command, which is otherwise the same as $DISPLAY R.

Three other system vectors are not automatically given values following a $FIT command but have to be extracted from the internal results of the fit. They are

%PE   Contains the estimates (length held in %PL).
%VC   Contains the variance-covariance matrix of the estimates (length held in %ML).
%VL   Contains the variances of the linear predictors.

To get at one or more of these type

$EXTRACT   vector name(s) $

listing the vectors you require.

10.7 Error messages and warnings

You will quite certainly have received numerous error messages by now, and these will have been accompanied by fairly detailed explanations of what has gone wrong and how it might be put right. You can switch these explanations off and on again by using the toggle directive

$HELP

If you switch help on immediately after a fault, you will get the explanatory message for that fault.

There are also some warning messages, some of which, like

CURRENT MODEL ABOLISHED

or

Model changed

in Glim77, you may have met. You can switch these off and on again by the toggle directive

$WARN

Switching off warnings is usually rather a rash thing to do.

10.8 The $OWN and $PASS facilities

If you really know your way around generalized linear models, you may find that the range of link functions and error distributions built into the package is inadequate to meet your needs. Glim allows you to specify your own link function and error distribution by writing four macros and specifying their names in an $OWN directive. The details are beyond the scope of this book; they are given in Chapter 8 of the Glim3 *Users' Manual* or Chapter 12 of the Glim77 *Users' Guide*.

@@@@
Some versions of Glim77 allow you to write a Fortran subroutine to manipulate Glim vectors. Macro texts can also be accessed. The subroutine must be called PASS and the $PASS directive is used to enter it. For details see Appendix F in the *Users' Guide*.
@@@@

Appendix A

The Glim directives

$ACcuracy 1.10	$FOrmat 1.5	$REStore 8.5
$ALias 4.5	*@$GRaph	$RETurn 8.2
$ARGument 9.3	@$Group 5.4	$REWind 8.5
@$ASSign 1.3	$Help 10.6	$SCale 2.4
$CAlculate 3.1	@$HIstogram 1.8	@$SEt 1.2
$C(omment) 1.2	$INput 8.2	$SKip 9.4
$CYcle 10.1	*@$LAyout	$Sort 3.6
$DAta 1.3	$LInk 7.5	$SSeed 3.5
$DElete 10.3	$Look 1.6	$STop 1.9
$DINput 1.4	$LSeed 3.5	$SUbfile 9.5
$Display App C	$Macro 9.2	$SUSpend 9.6
$(dummy) 1.2	*@$MANual	$SWitch 9.4
$DUmp 8.5	@$MAp 5.4	@$Tabulate 5.1
$ECho 8.2	$OFfset 6.2	**$Terms
$EDit 1.6	$OUtput 8.3	@$TPrint 8.6
$End 1.9	$OWn 10.7	@$TRanscript 1.2
$Endmac 9.2	$PAGe 8.4	$UNits 1.3
$ENVironment App C	@$PASs 10.7	$Use 9.2
$ERror App C	*@$PAuse	$Variable
$EXit 9.4	$Plot 1.7	$VErify 9.7
$EXTract 10.5	$PRint 8.6	$WArn 10.6
$FACtor 2.4	$Read 1.3	$Weight 4.2
$FINish 9.5	$RECycle 10.1	$WHile 9.4
$Fit 2.1	$REInput 9.5	$Yvariable

@   Not available in Glim3
*   Implementation dependent; may not be available
**   Ignored in Glim3; not allowed in Glim 77

The capital letters represent the minimum abbreviations.

Appendix B

Fortran-like conventions

B.1 Arithmetic expressions

A Glim $CALCULATE directive is followed by an *expression* made up of vectors and/or scalars and of *operators*. These are evaluated in the usual way; sub-expressions inside brackets are evaluated first, and apart from this the evaluation is controlled by the *priorities* of the operators. For the simple arithmetic operators these are

```
priority   1:   ** (exponentiation)
           2:   *, /
           3:   +, −
```

Sub-expressions involving operators of the same priority are evaluated from left to right, a rather confusing rule since it makes a/2*b mean (a/2)*b which may or may not be intended. Such constructions are best avoided.

All the arithmetic quantities in Glim are 'reals' in the Fortran or Pascal sense. They are handled by floating-point arithmetic, and this means that the results will usually not be exact. It is likely that ((3*1/3)==1) (or %EQ(3*(1/3),1)) will be false (equal to 0) because of rounding errors in the division.

If an expression does not take the form

```
name = sub-expression
```

its value is displayed; otherwise the value of the sub-expression is assigned to the entity named. Glim treats the assignment sign = as an operator with priority 7. Thus you can write

```
a = b*(c=(d+e))
```

The bracketed sub-expression (d+e) is evaluated and assigned to c *en passant* before the latter is multiplied by b. As a trick to get x displayed as well as assigned you can write

```
0+x=sub-expression
```

@@@@

In Glim77, expressions can contain the logical operators <, <=, ==, (is equal to), > =, >, /= (is not equal to), & (and), ? (or). Their precedences are

123

$<, <=, ==, >=, >, /=$: priority 4

& : priority 5

? : priority 6

@@@@

B.2 Format statements

A format defines the positions of data items in a *record*, one or more *lines* of data relating to a single unit. Each item occupies a *field*, a set of consecutive characters at a fixed position in the record. Each field is defined by a *format fragment* of the form Fw.d. Here w is a number giving the *width* of the field (i.e. the number of characters in it) and d is a number signifying the number of decimal places to be assumed. If an actual decimal point is one of the characters in a field, the value of d is ignored. Thus with fragment F4.2,

1234	means	12.34
-012	means	-0.12
34.5	means	34.50

If n consecutive fields have the same description, the fragment nFw.d can be used.

The other useful format fragment is nX which means 'skip n characters'. This can be used to skip over unwanted material in the record, which does not have to be numeric.

The slash symbol / in a format means 'start a new line'. If the fields defined up to the slash do not cover a whole line, the rest of the line is skipped.

As an example, suppose that a record consists of three lines of 80 characters each and that you only need a few items from the second line. These occupy character positions 1–5, 6–10, 25–30 and 31–33 in the record; the first three have integer values, the last is a percentage and is to be treated as if it had two decimal places. The format would be

(/2F5.0,14X,f6.0,f3.2/)

Appendix C

The $DISPLAY, $ENVIRONMENT, $ERROR and $LINK directives

It is convenient to bring together here some of the options available for these directives. For full details consult the Glim3 *Users' Manual* or the Glim77 *Users' Guide*.

C1 $DISPLAY options

C —the correlations of the parameter estimates.
D —the (scaled) deviance and its degrees of freedom.
E —the parameter estimates.
L —the make-up of the linear predictor.
M —the model details.
R —the y-values, fitted values and residuals.
S —the standard errors of estimate differences.
T —the g-inverse of the SSP matrix (with the y-variable).
V —the covariance matrix of the estimates.

C2 $ENVIRONMENT options

C —channel details
D —directory
I —implementation details
P —program-control stack
R —random number generator seed values
S —space allocation
U —usage of the data space and other structures

C3 $ERROR and $LINK options

The $ERROR options are

B —binomial
G —gamma
N —Normal
P —Poisson

The $LINK options are

C	—complementary log-log	$\ln\{-\ln(1-\mu/n)\}$
E	—exponential	μ ** (number)
G	—logit	$\ln\{\mu/(n-\mu)\}$
I	—identity	μ
L	—logarithm	$\ln(\mu)$
P	—probit	$NED(\mu/n)$
R	—reciprocal	$1/\mu$
S	—square root	$\sqrt{\mu}$

where n is the binomial denominator.

The default links are

for error	B	link	G
	G		R
	N		I
	P		L

One of C, G and P must be used with error B; they may not be used otherwise.

Index

127